AF553944

DEGRADATION BIOTECHNOLOGY

DEGRADATION BIOTECHNOLOGY

By

Dr. Ramesh K. Kothari

Department of Microbiology
Christ College,
Rajkot–360 005
Gujarat (India)

First Published-2005

ISBN 81-7141-986-0

Published by

DISCOVERY PUBLISHING HOUSE
4831/24, Ansari Road, Prahlad Street,
Darya Ganj, New Delhi-110002 (India)
Phone: 23279245 • Fax: 91-11-23253475
E-mail:dphtemp@indiatimes.com

Printed at:
Amit Enterprises, Delhi

DEDICATED

TO

THOSE WHO LOVES ENVIRONMENT

PREFACE

Environmental science research and industry is developing rapidly all over the world. Environmental science is increasingly being regarded as a core subject in most university, engineering branches and polytechnic biological sciences. There are already a number of excellent general textbooks on microbiology, biotechnology, environmental microbiology and environmental biotechnology in the market that deal with the basic principles of the field. In order to complement this, this book aims to focus on the various applications of microbial potential for bioremediation of different sites polluted with heterogeneous pollutants like synthetic textile dyes and poly-aromatic hydrocarbons by some bacteria and fungi. A "field applicable research based " is adopted whereby keeping in mind a case study. This book also includes information about the capability of microorganisms to cure nature itself, which can be exploit to clean up the contaminated sites consisting thousand types of biodegradable and recalcitrant organo-pollutants. This book will be useful particularly for the post-graduate students for reference who are interested to work in the field of environmental science. It may also serve as resource book for industries that release and treat solid as well as liquid wastes.

It is with great pleasure indeed that I dedicate the book to those who love science.

The book is divided into four chapters:

Chapter 1 provide a glimpse of introduction of subject and review of literature

Chapter 2 focuses with materials and methodology of various experiments carried out which include, isolation of various bacterial isolates, formulation of medium for isolation, cultivation condition for isolates, dye decolorization test, biomass vs dye decolorization, utilization of various organic compounds by bacteria, TLC and UV-Visible Spectra indicating decolorization of various textile dyes, dye degradation tests like, ring cleavage reaction, Gas Chromatography and Mass Spectroscopy (GCMS) and High Performance Liquid Chromatography (HPLC), degradation of aromatic amines.

Chapter 2 deals application of white rot fungus for bioremediation of contaminated sites, which include solid-state fermentation by WRF. Isolation of enzyme, crude enzyme preparation, dye decolorization assay like, Manganese Peroxidase (MnP), Manganese Independent Peroxidase (MIP), and Laccase as well as effects of veratryl alcohol and H_2O_2 on decolorization of various textile dyes.

Chapter 2 also deals exploitation of bacterial isolates for utilization of Poly Aromatic Hydrocarbons (PAHs) as a sole source of carbon.

Chapter 3 explores result profile of various experiments conducted at laboratory scale.

Chapter 4 provides a base for the open discussion on all the results including positive and negative with proper justification.

References, Plates and tables.

—Ramesh K. Kothari

ACKNOWLEDGEMENTS

"Interdependence is a higher value than independence"

This book is a synergistic product of many minds. I am grateful to the inspiration and wisdom of many scientists and far transgenerational sources. Many persons have contributed directly or indirectly to make this book a reality. I would like to express my gratitude towards them all.

I have immense pleasure in expressing my sense of gratitude to my esteemed guide Dr. S.J Pathak for his flawless guidance and systematic criticism during the present investigations. I am indebted to him for the many privileges and boundless compassion showered upon me throughout the research tenure.

I owe my arrearage and Dr. B.R.M Vyas for decisive suggestion, lending of literature and moral support during my tenure of work. In short, he is embodiment of honesty.

I am highly pleased to state that I could have not completed my work without the enthusiasm and the seed of the work that was sown by Dr. Pierre Cornelis, Department of Immunology, Parasitology and Ultra structure, VIB-Vrije Universiteit Brussel, Belgium.

I am highly thankful to Prof. V.C. Soni for his critical suggestions and moral support during my tenure of work.

I am also highly obliged to Prof. J. M. Dave and Prof. S.P. Singh, for their valuable remarks and co-operation throughout the period of my research work.

The support and help rendered by Dr. V.S. Rabadia, Dr. Vimal Bhuva, Mrs. Jigna Rabadia, Dr. Rajesh Patel, Dr. P.S.

Nagar, Mr. Shailesh Gondaliya at the beginning of my studies was of immense help for my research study.

I highly appreciate the help extended to me by Dr. K.S. Rajput, Ms. Hetal Bhimani, Mr Paresh joshi and Wilma Joshi linguistically peruse the entire book . I am heartily thankful to him.

I would also express affection to my wife Mrs. Charmy Kothari and my beloved son Master Dhyey Kothari for their unfailing forbearance, encouragement and affection during the worst of worst situation and patience during preparation of this book.

Thanks are also due to all those who have not been mentioned by name but nevertheless have been of invaluable help in their inscrutable ways.

—Ramesh Kothari

CONTENTS

LIST OF ABBREVIATIONS

CM	Complete Medium
CMB	Complete Medium Broth
CMA	Complete Medium Agar
MM	Mineral Medium
CETP	Common Effluents Treatment Plant
UT-1	Sample site-1 for untreated sample
UT-2	Sample site-2 for untreated sample
SF	Sludge Free
S	Sludge
T	Treated
gl^{-1}	Grams per liters
RPM	Revolution Per Hours
Mg	Milligrams
Ml	Milliliter
EM	ErlemMeyer Flask
Jd	Dye collected from Jetpur textile industries
Ad	Dyes collected from ATUL dyestuff industries
Kd	Dyes collected from Karnavati (Ahmedabad) ASHIMA dyes stuff
Pseu-I	Pseudomonas sp. Strain I
Pseu-II	Pseudomonas sp. Strain II

Pseu-III	Pseudomonas sp. Strain III
MGD	Minerals + Glucose + Dye
MYED	Minerals + Yeast extract + Dye
MPD	Minerals + Peptone + Dye
MBAD	Minerals + Benzoic acid + Dye
G-1	Glucose + Peptone + Minerals + Dye
G-2	Glucose + Yeast extract + Minerals + Dye
G-3	Yeast extract + Peptone + Minerals + Dye
P-1	Benzoic acid + Peptone + Minerals + Dye
P-2	Benzoic acid + Yeast extract + Minerals + Dye
OD	Optical density
CFU	Colony Forming Unit
Cm	Medium as a control
Cd	Dye as a control
T	Decolorized (*i.e.* tested) dyes
TLC	Thin Layer Chromatography
D	Dye
M	Medium only
MO	Medium + Organism
MDO	Medium + Dye + Organisms
M	Molar
MM	Milli Molar
μg	Micrograms
DEE	Diethyl ether
Rf	Relative factor
MSM	Mineral Salt Medium

SLB	Succinate Lactate Buffer
TB	Tartrate Buffer
CEEP	Concentrated Extracellular Enzyme Preparation
ES	Enzyme Sample
LMEs	Lignin Mineralizing Enzymes
WRF	White Rot Fungi
PAHs	Poly Aromatic Hydrocarbons
MnP	Manganese Peroxidase
MIP	Manganese Independent Peroxidase
RS	Reactofix Supra
BOD	Biological Oxygen Demand
COD	Chemical Oxygen Demand
FMEM	Fortified Malt Extract agar Medium
SSF	Solid State Fermentation
Cp	*Coriolopsis polysona*
Rc	Newly isolated white rot fungus sp. *Strain Rc.*

Key words

Biodegradation

Bioremediation

Bioaugmentation

Polyaromatic hydrocarbons (PAHs)

Biotransformation

Bioaccumulation

Recalcitrant

Xenobiotics

Textile dyes

Azo dyes

Lignin

Organopollutants

Pseudomonas sp. Strain-II

Coriolopsis polysona

Rc (Newly isolated white rot fungus-Basidiomycetes)

One of the tools for a better and cleaner tomorrow is in our hands. We must work with that tool, towards understanding its limitations and its horizons. Our environment, the combination of soil, air and water, gave birth to life, as we know it. We have taxed its ability to absorb the wastes we create. We must work now to clean it with the same organisms from which life is supposed to have come.

1

AN INTRODUCTION

AN OVERVIEW

The past decades have witnessed an increasing awareness that human activities, in particular intensive agriculture and industrial technologies, must be brought in harmony with the global material cycles in the biosphere. In other words, we shall have to make a transition from exploitation of our natural resources towards a partnership with the global ecosystem. Although in recent years much emphasis has been placed on the need to reduce the contamination of soils, ground water, sediments, surface water and air with hazardous chemicals, the issue is in fact much broader. The technologies of the more distant future should be based on renewable resources, they should use "mild" production processes, the resulting products and services should be environmentally compatible and any waste they generate should be recyclable (OECD, 1994).

Since before the beginning of the century, human beings have dumped enormous amounts of waste products into the environment following the principle of "*out of sight, out of mind*". Until World War II (WWII), these waste products and their effects went largely unnoticed. However, after WWII, the severity of pollution problem, resulting from careless waste disposal, has steadily increased. Historically, sources of waste products have been industrial and agricultural. For instance, the energy production industry and textile printing and dyeing industries generates huge amount of waste during the processing. In past, these wastes have been buried and released in open environment (in stagnant water bodies as well as flowing water streams bodies), some times not so carefully, and as a result waste constituents have migrated through the soil into ground water bodies forming a point source of contamination. On the other hand, the agricultural sector is the nation's largest user of biocides and fertilizers. The application of biocides and fertilizers over vast land areas is responsible for the source of nonpoint source of contamination (Raina *et al.*, 2000).

We as a society are reexamining the cleanup issue from several perspectives. The first is related to the clean up target and the question "*How clean is clean*". Society began responding to environmental concerns long ago; beginning with the recognition that our environment is fragile and human activities can have a great impact on it (Raina *et al.*, 2000).

With the advent of manufacture of synthetic chemicals, many potentially hazardous organic compounds have been introduced into various components of the environments. One method for removing these undesirable compounds from the environment is bioremediation—application of biological treatment to the cleanup hazardous chemicals, an extension of carbon cycling (Sutherland *et al.*, 1999).

BASIC UNDERSTANDING

Biodegradation, mineralization, bioremediation, bio-deterioration, biotransformation, co-metabolism, and bio-accumulation are terms not always used with appropriate sensitivity to their subtle differences in meaning.

Biodegradation Concept and Process

Decomposition is a natural process that has been established since the origin of the earth. Living organisms, after origin and evolution, developed processes of utilization of molecules that either synthesized by them or existed due to physico-chemical reactions on the earth's environment. Thus, the recycling of elements and/or molecules have been the part and partial of evolution of life.

In the process of decomposition, biotic components depolymerized, disintegrate or directly assimilate and metabolize several biomolecules produced by them. Thus, the process does not allow accumulation of dead biomass and that is being recycled.

Decomposition process includes the biodegradation or both the terms are synonyms and may be employed alternatively.

Microbes and Biodegradation

Microorganisms are among the main agents that perform the process of biogeochemical cycling.

Biodegradation : It is an umbrella term, encompassing most of the other jargon. It involves chemical transformations mediated by microorganisms that satisfy nutritional and energy requirements, detoxify the immediate environment or occurs fortuitously such that the organism receives no nutritional or energy benefit (Stoner, 1994).

Mineralization : It is the complete biodegradation of organic materials to inorganic products, and often occurs though the combined activities of microbial community rather than a single organism (Shelton and Tiedje, 1994).

Co-metabolism : It is partial biodegradation of organic compounds that occurs fortuitously and that does not provide energy or biomass to the organism(s). Co-metabolism can result in partial transformation to an intermediate that can serve as a carbon and energy source for organisms. Thus, when an existing enzyme happens to have catalytic activity towards a compound, which is not its normal substrate (may or may not yield energy or a dead end intermediate (Horvath, 1972; Gibson, 1991, and Hulbert and Krawies, 1977).

Biotransformation : It is the conversion of an organic compound to another, which is not complete, *i.e.,* does not result in the formation of only inorganic products or cell mass.

Bioremediation : It is a technology developed employing biological system(s) to transform and/or degrade undesirable (it may be toxic) compounds or otherwise render them harmless, either *in situ* or *ex situ*. In bioaugmentation, a process of bioremediation, in which exogenous biological system(s) are, added (Christon *et al.*, 2002).

Xenobiotics : Xeno, means foreign to nature, are synthetic compounds that do not occur naturally, *i.e.*, man-made or may be called anthropogenic compounds.

Biodegradability is a key property of the environment, and thus, any substance entering into the environment may undergo variety of biodegradation processes. In present day, thousands of chemicals, which are produced or utilized for domestic, industrial, or agricultural purposes and discharged, ultimately entering into any component of the environment. These compounds may further be degraded or transferred to the other environmental components. However, many xenobiotic compounds which are not susceptible to biotic action and thus, may accumulate and create hazards in the environment.

BACTERIA AND TEXTILE DYES

The history of the beautiful colored compounds-dyes, which makes our life full with colors and sometimes dark with death, goes back over hundreds of years. Dyes have been used since ancient times for coloring and printing fabrics. In these periods only the kings and noblemen could afford color fabrics. A dye dress was a symbol of riches.

William Henry Perkin, as a young eighteen-year-old student of Hofmann, synthesized the first dye, Mauva, by accident in 1856 from chemicals derived from coal. He decided to make quinine from aniline. Instead of quinine, he obtained a black mass in his reaction. On working up, he found that he had made a purple dye. The success of Perkin later on led to accelerate work in this field, which resulted in establishing dyestuff industries on a large scale in Europe. Several hundred varieties of dyes are produced on a large scale in our country today. Starting in a small way after the Second World War, the dyestuff industries today export dyes to other countries. This is the beginning of a chemical industry producing approximately

700,000 tons of colorants on yearly bases, 131 years later (Zollinger, 1987). It is very difficult to get recent estimated data on the worldwide production of dyes. However, it can be assumed that more colorants are produced in view of the economic prosperity and also because the production of many colored goods increased and color printing, dyeing and copying became normal currently in every field.

Industrialization is commonly believed to be the universal remedy for trade and industry backwardness. Mounting pressure on industrialization to withstand in the context of advancement towards economic stability is constantly degrading the environment through air, water and soil pollution (Rajannan and Oblisami, 1979; Slobe, 1986; Mahnot and Bhandari, 1987).

Most of the dyes made in 19th century were derived from the aromatic intermediate compounds obtained from the coal tar distillation through the dyes industry originated in England, but the rapid growth occurred in Germany. The textile industry is one of the largest, never ending, indispensable and a part of backbone industries of some under developed and developing countries in the world. It is classified into cotton, woolen, and synthetic fibre sectors. The pretreatment, the process of coloring, and after treatment of these fibers (washing) usually require large amounts of water and a variety of other chemicals.

DYESTUFF INDUSTRY IN INDIA

The dyestuff industry in India started around 1940 by establishment of a small plant by MS Associated Research Laboratories. They manufactured rapid fast dyes. The Atul Products Ltd. was the first major dye producing plant established in 1952. Later on several manufacturers such as Indian Dyestuff Industries, Amar Dye Chem., Color Chem., Sudarshan Chemicals and a large number of smaller units started manufacturing dyes. In most developed countries a few well-established companies synthesize dyes while in India dyes are manufactures by a large number of small companies along with a few large companies producing almost full range of dyes.

Some Groups of Dyes and Dyeing

Direct Dyes

These are mostly azo dyes dyed on cotton form a water solution at 50-100°C, and these have moderate to poor fastness.

Sulphur Dyes

These are mostly used for cotton from a dye bath containing Sodium Sulphide. This dye has good fastness properties and gives black, green, blue and brown shades.

Azoic Dyes

These are two component systems, which produce an insoluble azo pigment on the fibre. The fibre to first impregnate in a coupling component squeezed uniformly and dipped in a diazosolution when the insoluble dye is formed. These give bright shades especially in yellow, orange and red, with good fastness.

Ingrain Dyes

Basically these are similar to azoic dyes in the sense that the dye is formed on the fabric form intermediate compounds. This group consists of phthalogen dyes, which give phthalocyanine pigments on the fabric.

Vat Dyes

This is a very large and important group of dyes, which is applied, in soluble form in alkaline hydrosulfite solution. On keeping the fibre in air the leuco dye is oxidized into insoluble form. Almost all shades with good fastness properties are available in this group. These dyes are used for cotton, wool, silk and cellulose acetate.

Acid Dyes

These dyes are mostly used for wool but sometimes find application on silk, polyamide acrylic and other fibres as well as paper and leather. The dyeing is carried out normally under acidic condition. These dyes contain all shades of colour. The fastness properties also are in a vide range from poor to good.

Mordant Dyes

These dyes by themselves have poor affinity for the fibre but form complexes with metals. The complexes are many times referred to as lakes. Many acid dyes can be applied to cotton with the help of metals such as copper and chromium.

Metal Complex Dyes

These are mostly used for wool. They consist of a metal complex of an azo dye. They are dyed from acidic or neutral solution and have good fastness properties.

Dispenser Dyes

These are water insoluble dyes applied from fine aqueous suspension and are used for dyeing cellulose acetate, nylon, polyester, and polyacrylonitrilet fibres. The dyeing involves dissolution of the dye in the fibre.

Pigments

In the last decade increasing use of insoluble stable pigment is made for imparting color to the fabrics. The pigment is suspended in an emulsion of polymers. On application and drying the pigment gets bound to the polymer sticking to the fabric. Pigments can also be incorporated into synthetic fibres before spinning.

Reactive Dyes

These dyes contain a reactive group, which reacts with –OH group of cellulose to form a covalent linkage with the fibre.

Basic Dyes

They dye cotton and give intense and brilliant shades but have poor light fastness.

Solvent Dyes

These dyes are a soluble in organic solvents and are used for manufacture the stains, varnishes, inks, lacquers, copying papers, typewriter ribbons, candles, polishes, soap, cosmetic etc. Dyes are most commonly used for application on cellulosic, protein, polyamide, polyesters and polyacrylonitrilet

Textile dyes are widely used in a number of processes, such as textile dyeing, paper printing and color photography. At present over 100,000 dyes are commercially available with 7×10^5 tons of dyestuff being produced annually (Meyer, 1981).

Azo dyes are by far the largest and most important group of dyes (Carliell *et al.*, 1998). The fundamental of the production of azo dyes laid in 1858 when P. Gries discovered the reaction mechanism, diazotization, for the production of azo compounds (Zollinger, 1987). Nowadays azo dyes are used for coloring many different materials such as textile, plastics, food and food packing, leather and its ornamental preparations, pharmaceuticals and for manufacturing paints and lacquers as well as for printing purposes in each every fields. Azo dyes absorb light in visible range spectrum range and this property is attributed to its chemical structure. Their chemical structure is characterized by one or more azo groups (–N=N–). The azo group is substituted with benzene or naphthalene groups, which can contain many different substituents such as Chloro (–Cl), Methyl (CH_3), Nitro (–NO_2), Amino (–NH_2), Hydroxyl (–OH) and Carboxyl (–COOH). A substituents often found in azo dyes is the sulfonic acid group (–SO_3H).

Many of the intermediates and dyes have high toxicity and the workers in the industry are affected. Workers hesitate to work in dye industries and hence the dyestuff as well as some intermediates is being exported from India. We still make a large number of direct dyes from benzidine. Most of the dyes and intermediates plants in India are located in Gujarat and Maharastra. This is due to the early establishment of textile and chemical industries in these states.

Dye manufacture requires several other chemicals, for making one kilogram of a dye 6-10 kilogram of other basic chemicals are required. The requirements of our dyestuff industries for basic primary organic chemicals are Acetic acid, Acetic anhydride, Benzene, Methanol, Naphthalene, Toluene, Urea, Xylene etc., inorganic chemicals like, Aluminium chloride, Ammonia, Caustic potash, Caustic soda, Chlorine, Chlorosulphonic acid, Hydrated lime, Hydrochloric acid, Iron powder, Nitric acid, Soda ash, Sodium bisulfite, Sodium Nitrite, Sodium Sulphide, Sulphuric acid, Zinc dust etc. are used for the manufacturing of azo dyes, metal azo dye complexes, direct dyes, fast dyes, sulphur dyes, pigment emulsions, vat dyes, disperse dyes, reactive dyes, ingrain dyes, optical brighteners, basic dyes, acid dyes, solvent dyes, mordant dyes etc.

DYES AND ENVIRONMENTAL PROBLEMS

Contamination of wastewater generating from the industries such as dye manufacturing plants and dyeing and printing processing houses are of great environmental concern. The wastewaters arise from the industries containing dye intermediates and dyes are difficult to remove by the conventional effluent treatment processes available. Thus, causes pollution due to discharge of large volume of untreated (or partially treated) wastewaters, our environment and/or their components. Due to, inefficiency in the dyeing process results in 10-15% of all dyestuff being directly lost to wastewater, which will ultimately find its way into the environment (Vaidya and Datye, 1982; Zollinger, 1987). It is likely this figure is presently even higher since the usage of reactive dyes (which include azo dyes) increased lately and their fixation rate in dyeing processes can be as low as 50% (Easton, 1995).

There exist a clear and strong relationship between the chemical structure of dyes and their recalcitrant nature. It seems that some of the dyes when degraded biotically or abiotically, produce the end products that are more toxic than the native dyes. Examples of such harmful moieties are 1,4-phenylenediamine, 1-amino-2-naphthol, benzidine and substituted benzidine, like o-toluedine (Chung *et al.*, 1981; Reid *et al.*, 1984; Rosenkranz and Klopman, 1989; and Rosenkranz and Klopman, 1990). Some azo dyes are non-mutagenic, their reductive degradation products, particularly, aromatic amines, are mutagenic (Chung *et al.,* 1981).

The benzidine moieties in azo dyes are presently prohibited, benzidine analogue dyes. Evidence clearly indicate that sulfonated azo dyes show decreased or no mutagenic effect compared to unsulfonated azo dyes due to their electric charge and low lipophilicity, which prevents them from uptake and metabolic activation (Chung and Cerniglia, 1992; Jung *et al.,* 1992; Levine, 1991 and Rosenkranz and Klopman, 1989).

Certain dyes contain copper or other metals as an integral part of the dye molecule. A review of the dye structure in the color index indicates that most of these metal-containing dyes are either blue or green. There are many of these dyes, mostly in the 74000 series of chemical constitutions, which are phthalocyanine dyes and pigments. For example, Vat Blue 29 which contain cobalt, Pigment Blue 15 (Copper), Ingrain Blue 14 (Nickel), Ingrain Blue 5 (Cobalt), Ingrain Blue 13 (Copper), Direct Blue 86 (Copper), Direct Blue 87 (Copper), Pigment Blue 17 (Copper, Barium), Acid Blue 249 (Copper), Ingrain Blue 1 (Copper), Pigment Blue 15 (Copper), Pigment Green 37 (Copper), Pigment Green 7 (Copper), Ingrain Green 3 (Copper), Solvent Blue 25 (Copper), Solvent Blue 24 (Copper), Solvent Blue 55 (Copper), Reactive Blue 7 (Copper), Brent Smith (1988). Due to the above-mentioned effects, it is clear that azo dyes should not enter into the environment without proper treatment. Out of the attractive method, to prevent this is to apply microbial treatment methods for their mineralization. Textile mill effluents are complex mixtures of chemicals, varying in composition over time and from unit to unit. Untreated effluents from textile units have been characterized as having extreme pH, elevated temperature, and high load of Chemical Oxygen Demand (COD), Total Suspended Solids (TSS), oil, grease, and metals.

Solids in textile wastewater are from process wastewater generated from fibrous substrate and process chemicals, and from biological wastewater treatment. The solids disturb the aquatic system by slowing oxygen transfer and reducing light penetration. They can cause the formation of anaerobic sludge layers in riverbeds.

Some processes, such as dyeing of wool with acid leveling dyes, are highly acidic while others, such as dyeing of cotton with reactive dyes are highly alkaline. As a result, wastewater pH can vary greatly over a period of time and depending on the processes performed in a particular textile factory. Thus neutralization is necessary. The extent of neutralization depends on the extreme of the pH and the alkalinity-acidity of the wastewater.

In principle, decolorization is achievable using one or a combination of the following methods: adsorption, filtration, precipitation, chemical degradation, photo degradation, and biodegradation. However, the color and the chemical composition of the textile effluents are usually subjected to both the daily and seasonal variations dictated by production routines and fashion cycles. A single, universally applicable end-of-pipe solution is therefore implausible

Many synthetic dyes such as azo dyes, are resistant to microbial degradation under the aerobic conditions normally found in wastewater treatment plants. This is because dyestuffs are designed to be resistant too chemical fading and light induced oxidative fading. Other factors involved in reducing the biodegradation of dyes include properties such as high water solubility and high molecular weight, which inhibit permeation though biological cell membranes. Many dyestuffs, in particular disperse, direct and basic dyes, are expected to be removed from wastewaters via adsorption onto activated sludge. Acid dyes and reactive dyes, however, exhibit low adsorption values and thus pass though activated sludge processes largely unaffected.

Most studies on azo dyes biodegradation have focused on bacteria (Horitsu *et al.*, 1977; Edaka *et al.*, 1980, 1982; Brown, 1981; Zimmermann *et al.*, 1984; Lin and Yang, 1989, 1991, 1992; Yatome *et al.*, 1993; Rafii *et al.*, 1990; Cao *et al.*, 1993; Govindaswami *et al.*, 1993; Nigam *et al.*, 1996b), reported the decolorization of effluent from the textile industry by Alcoligens faecalis, Commomonas acidovorans.

Tan et. al., (2000) reported biodegradation of azo dyes namely 4-phenylazophenol (4-PAP) and Mordant Yellow 10 (4-

Sulfophenylazo-Salicylic acid; MY-10) in Co-cultures of anaerobic granular sludge with aerobic amine degrading enrichment cultures.

Decolorization and biodegradation of N, N[1]-dimethyl-p-phenylenediamine-an aromatic amines by Klebsiella pheumoniae RS-13 and Acetobacter liquefaciens S-1 was studied by (Wong and Yuen, 1996).

Synthetic dyes such as a 20 dyes, Xanthene dyes and anthraquinone dyes are very toxic to living organisms (Gingell and Walker, 1971; Mayer, 1981; Manning *et al.*, 1985; Rafii *et al.*, 1990).

In most biological base treatment systems the possibility exists of enriching the microbial population with specifically isolated bacterial species to aid degradation of a particular dyestuff or dye class. Examples of acid dyes and reactive azo dyes have been found to be degraded by specific bacterial species.

Several waste treatment systems based upon aerobic and anaerobic bacterial action have been developed to treat textile industry wastewater. Currently, interest is concentrated on the removal of amines which are the degraded products of some dyes by indigenous microorganisms, heavy metals and the operation of biological effluent treatment systems in much more complex and toxic effluent streams.

A considerable effort has been spent on developing suitable treatment systems for these effluents. Only biotechnological solution can offer complete destruction of the dyestuff, with a reduction in COD. In addition, the biotechnological approach makes efficient use of the limited development space available in many traditional dyestuffs.

ANAEROBIC BIODEGRADATION OF AZO DYES

Under anaerobic conditions azo dyes are readily decolorized, as a result of the reductive transformation of the azo group. Due to this reduction two aromatic amines are formed the aromatic amines do not absorb light in the visible spectrum and therefore azo dye reduction represent a decolorization process. The anaerobic decolorization of azo dyes was first investigated using intestinal anaerobic bacteria (Allan and Roxon, 1974; Brown, 1981; Chung and Cerniglia, 1992; Walker and Ryan, 1971). Later, these compounds were found to become also readily decolorized with various other anaerobic cultures (Beydilli *et al.*, 1998; Brown and Laboureur, 1983b; Carliell *et al.*, 1995; Donlon *et al.*, 1997; Razo-Flores *et al.*, 1997). The exact mechanism of the anaerobic azo dye reduction is not clearly understood yet. Therefore, the term azo dye reduction may involve different mechanisms or locations like enzymatic (Haug *et al.*, 1991; Rafii *et al.*, 1990), non-enzymatic (Gingell and Walker, 1971; Kudlich *et al.*, 1997; Van der Zee *et al.*, 2000a), intracellular (Mechsner and Wuhrmann, 1982, Wuhrmann *et al.*, 1980), extracellular (Carliell *et al.*, 1995) and various combinations of these mechanisms and locations.

In one case even a complete anaerobic mineralization of the azo dye—azodisalicylate was observed under methanogenic conditions. This azo dye was first reduced to the aromatic amine 5-aminosalicylic acid (5-ASA) and next 5-ASA was anaerobically mineralized (Razo-Flores *et al.*, 1997).

A precondition for the reduction of azo dyes is the presence and availability of a co-substrate (Nigam *et al.*, 1996b), because it acts as an electron donor for the azo dye reduction. Many different co-substrates were found to suite as electron donor, like glucose (Carliell *et al.*, 1995; Nigam *et al.*, 1996b), hydrolyzed starch (Willetts *et al.*, 2000), tapioca (Chinwekitvanich *et al.*, 2000), Yeast extract (Nigam *et al.*, 1996a), a mixture of acetate, butyrate and propionate (Donlon *et al.*, 1997; Van der Zee *et al.*, 2000c) and the azo dyes reduction product 5ASA as well (Razo-Flores *et al.*, 1997).

It also has been observed that the extend of decolorization of an azo dye like Remazol Black B varies depending on the co-substrate used, *e.g.*, 82% with glucose, 71% with glycerol and lactose, 51% for starch and 39% for a distillery waste (Nigam *et al.* 1996b). Moreover, also the rate of azo-reduction process depends on the type of co-substrate used and/or on the chemical structure of the azo dyes. (Van der Zee *et al.*, 2000b; Van der

Zee *et al.*, 2000c). Furthermore, it has been observed that compounds that facilitate the transport of electrons, like mediators, considerably enhance the azo reduction (Kudlich *et al.*, 1997; Van der Zee *et al.* 2000b). This suggests the prevalence of an extracellular process.

Even some biodegradation of azo dyes can act as mediator (Keck *et al.*, 1997; Van der Zee *et al.*, 2000c). The reduction of azo dyes proceeds better under anaerobic thermophillic conditions than under mesophilic conditions, although the thermophillic process to be less stable compared to the mesophilic process (Willetts *et al.*, 2000).

Microorganisms are known to play a crucial role in the mineralization of xenobiotic compounds, like azo dyes (Lie *et al.*, 1998). The mineralization of azo dyes requires an integrated or sequential anaerobic and aerobic steps (Field *et al.*, 1995). The first step in the biodegradation of azo dyes concerns the azo dye reduction that readily proceeds under anaerobic conditions and results in the formation of aromatic amines (Carliell *et al.*, 1995; Razo-Flores *et al.*, 1997; Walker, 1970; Weber and Wolfe, 1987). Anaerobic consortia generally do not degrade the aromatic amines but most of the aromatic amines are readily biodegraded under aerobic conditions (Baird *et al.*, 1977; Brown and Laboureur, 1983a).

AEROBIC BIODEGRADATION OF AZO DYES

Under aerobic condition, completely degradation of azo dyes is not possible and many times it remains recalcitrant. Some specific aerobic bacterial cultures were found to have ability to reduce the azo linkage via an enzymatic reaction. The azo reductases isolated from these organisms have a narrow substrate range (Kulla 1981; Kulla *et al.*, 1983; Zimmermann *et al.*, 1984; Zimmermann *et al.*, 1982a; Zimmermann *et al.*, 1982b).

The occurrence of aerobic conversions of sulfonated azo dyes were reported by (Heiss *et al.*, 1992 and Shaul *et al.*, 1991), and sometimes even a complete mineralization of a sulfonated azo dye was found under aerobic conditions. A bacterial strain S5 was able to reduce the azo dye 4-carboxy-4'-sulfobenzene and to mineralize the azo dye reduction products 4'-aminobenzenesulfonic acid (4-ABS) and 4-aminobenzoic acids. The sulfonated azo dye was used as carbon and energy source in this case (Blumel *et al.*, 1998).

Also a bacterial strain MI-2 isolated from a biofilm reactor, was able to utilize Acid Orange 7 and 8 as sole source of carbon, nitrogen and energy source and the azo dye reduction product 4-ABS was also degraded (Coughlin *et al.*, 1997).

AEROBIC BIODEGRADATION OF AROMATIC AMINES

It has been extensively studied that many aromatic amines can undergo degradation under aerobic condition. Many of these compounds were found to be degraded under aerobic conditions (Baird *et al.* 1977; Brown and Laboureur, 1983a), *e.g.*, compounds like aniline (Anson and MacKinnon, 1984; Konopka 1993; Loidi *et al.*, 1990; Lyons *et al.* 1984; Patil and Shinde, 1988), carboxylated aromatic amines (Russ *et al.* 1994; Stolz *et al.*, 1992), chlorinated aromatic amines (Hwang *et al.*, 1987; Loidi *et al.*, 1990; Reber *et al.*, 1979) and (substituted) benzidine (Baird *et al.*, 1977). Many time some aromatic amines cannot undergo biodegradation are sulfonated aromatic amines (Bretscher, 1981; Tan and Field, 2000). These aromatic amines are released into the environment mainly via either the production or the biodegradation of sulfonated azo dyes.

The ability of aerobic bacteria to mineralize sulfonated aromatic compounds was observed for the first time in the seventies. As a result of this biodegradation, the sulfur moiety of the sulfonic acid group can enter in the sulfur cycle (Cain and Farr, 1968; Focht and William, 1970; Ripin *et al.*, 1971).

Their fate of alkylbenzenesulfonic acids when released in the environment has been extensively studied and reviewed (Berna *et al.* 1991; Federle and Ventullo, 1990; Jimenez *et al.*, 1991; Mampel *et al.*, 1998, and Thoumelin, 1991).

The group of aromatic sulfonates comprises the sulfonated azo dyes themselves and their precursors and biodegradation products the sulfonated aromatic amines and some additional sulfonated aromatic compounds.

Aromatic sulfonates present the raw materials for the production of azo dyes, drugs, detergents, optical brighteners, and artificial sweeteners (Hansen *et al.*, 1992; Locher *et al.*, 1989). Some aromatic sulfonates have a direct applicability *e.g.*, 4-ABS is used as a preservative (Hooper, 1991), 3-ABS as a mild oxidants (Locher *et al.*, 1989) and 2- benzo sulfonic acid (2-BOS) as a wetting agent in toothpaste (Hansen *et al.*, 1992).

The recent high profile of color pollution is mainly the result of increasing public awareness and expectations of the environment, coinciding with rising levels of colored discharge and due to the above-mentioned effects, it is clear that dyes should not enter in the environment. An attractive method to prevent this, is to apply microbial treatment methods for their mineralization.

BACTERIA AND POLYAROMATIC HYDROCARBONS (PAHs)

Polycyclic aromatic hydrocarbons (PAHs) are widely distributed environmental contaminants that have detrimental biological effects, including acute and chronic toxicity, mutagenicity, and carcinogenicity (Cerniglia, 1992; Gibson and Subramanian, 1984; Wild *et al.*, 1990 and Dipple *et al.*, 1990).

PAHs are a class of compounds consisting of fused aromatic ring in various structural configurations. PAHs with four and five fused benzene rings are more resistant to biodegradation than PAHs with two or three rings.

PAHs and its Environmental Significance

Due to the hydrophobicity, low water solubility and electrochemical stability, PAHs compounds are highly persistence in the environment. Higher molecular weight PAHs exhibit greater environmental persistence than lower molecular weight PAHs due to an increase in hydrophobicity and an increase in resonance stabilization with increasing mass. This relationship of environmental persistence and increasing number of benzene rings is consistent with the results of studied correlating biodegradation rates and PAHs molecule size (Bossert and Bartha, 1986; Heitkamp and Cerniglia, 1987).

Source of environmental PAHs include power plants, domestic heating systems, which burn oil, coal or wood for example, gasoline and diesel engines, waste incineration, various industrial activities, and tobacco smokes (Hall and Grover, 1990). More specifically, petroleum-refining processes contribute to localize loadings of PAHs into the environment through industrial effluents from coal gasification and liquefaction processes and accidental spillage of raw and refined petroleum (IARC, 1972-1990).

PAHs are hydrophobic compounds whose persistence within ecosystems is chiefly due to their low aqueous solubility, which makes them less available for biological uptake (Futoma *et al.*, 1981; Landrum *et al.*, 1994; Van der *et al.*, 1994 and Wild *et al.*, 1990).

Due to the ubiquitous occurrence, recalcitrance, bioaccumulation potential, and carcinogenic activity of PAHs, many researchers have developed procedures for the biodegradation of PAHs contaminated sites (Carmichael *et al.*, 1997; Carmichael and Pfaender, 1997; Catallo and Portier, 1992; Cerniglia and Heitkamp, 1990; Ghisalba *et al.*, 1985; Guthrie and Pfaender, 1998 and Stringfellow and Aitken, 1994).

At present, many microorganisms have been shown to metabolize the three-ring PAHs, phenanthrene and anthracene. The four-ring PAHs, pyrene, chrysene and Benz[a] anthracene have also now been shown to be susceptible to microbial biodegradation (Sutherland *et al.*, 1995; Cerniglia, 1992).

Some microorganisms can oxidize these chemically stable PAHs when grown on an alternative carbon source, the process is known as co-metabolism.

PAHs Degradation by Bacteria

Generally, soils with a high total PAHs content contain more PAHs utilizer (10^5-10^{10} bacteria per gram of soil) than soil with low PAHs content. A high number of PAHs degraders also correlate well with the amount of biodegradation activity in soil (Shiaris, 1989). *Pseudomonas* sp. (Muller *et al.*, 1990; Weissenfels *et al.*, 1990; Grifoll *et al.*, 1992), *Alcaligenes* sp. (Weissenfels *et al.* 1991), *Rhodococcus* sp. (Walter *et al.*, 1991), *Beijerinckia* sp. (Mahaffey *et al.*, 1988), *Mycobacterium* sp. (Kelley *et al.*, 1993; Kelly and Cerniglia, 1991; Grosser *et al.*, 1991 and Heitkamp *et al.*, 1988), *Staphylococcus* sp. (Monna et at., 1991), *Arthrobacter* sp. (Efroymson *et al.*, 1991; Keuth and Rehm, 1991) have been isolated that can completely degrade or co-metabolize higher molecular weight PAHs, such as fluoranthrene, pyrene, fluorine and benzo[a]anthracene.

Proposed pathways for the biodegradation of these PAHs have also been elucidated. The bacterial degradation of PAHs initially proceeds through a dioxygenase attach on an aromatic ring to form a cis-dihydrodiol. Monooxygenase oxidation to form trans-dihydrodiols has also been shown (Heitkamp and Cerniglia, 1987; Kelley and Cerniglia, 1991). The degradation of naphthalene, the simplest PAHs, has been studied in great detail (Gibson, 1984). The enzymatic mechanisms and genetic regulation of the plasmid-encoded naphthalene catabolic pathway of *Pseudomonas* has been ell characterized (Zylstra *et al.*, 1991).

Aerobic biodegradation of the two and three-ring PAHs is accomplished by number of soil bacteria. As the number of fused rings and the complexity of the substituted groups increase, the relative degree of degradation decreases. The influence of alkyl substituents is more difficult to predict (Cookson, 1994). One methyl addition significantly decreases the degree of degradation, and its effect varies with the substituted position. The addition of three methyl groups causes severe retardation of degradation (McKenna, 1979). The co-metabolism for PAHs having four or more rings has been demonstrated by several investigations. In fact, co-metabolism may be the only metabolism mode for degradation of the heavier PAHs. Presence of an analog substrate such as naphthalene will enhance the degradation of pyrene by many microorganisms (Cookson, 1994).

Prior to the degradation of many organic compounds, a period is observed in which no degradation of the chemicals is evident. This time interval is known as the acclimatization period or, sometimes as adaptation or lag period. The length of the acclimatization period varies and may be less than 1 h or many months. The duration of acclimatization depends upon the chemical structure, biogeochemical environmental conditions, and concentration of the compound.

Metabolic Modes

The design of bioremediation processes requires determination of the desired degradation reactions to which the target compounds will be subjected. This involves the selecting the metabolic mode that will occur in the process. The metabolism modes are broadly classified as aerobic and anaerobic. Aerobic transformation occurs in the presence of molecular oxygen, with molecular oxygen serving as the electron acceptor. This form of metabolism is known as aerobic respiration. Anaerobic reactions occur only in the absence of molecular oxygen and the reaction is subdivided into anaerobic respiration, fermentation, and methane fermentation.

Microorganisms have developed a wide variety of respiration systems. These can be characterized by the nature of the reductant and oxidant. In all cases of aerobic respiration, the electron acceptor is molecular oxygen. Anaerobic respiration uses an oxidized inorganic or organic compound other than oxygen as the electron acceptor. The respiration of organic substrate by bacteria is, in most cases, very similar. The substrate is oxidized to CO_2 and H_2O.

Co-metabolism

Another important metabolism concept in bioremediation is co-metabolism. In a true sense, co-metabolism is not metabolism (energy yielding), but unexpected transformation of a compound. It was a traditional belief that microorganisms must obtain energy from an organic compound to biodegrade it. The transformation of an organic compound by a microorganism that is unable to use the substrate as a source of energy is termed co-metabolism (Alexander, M., 1994).

Enzymes generated by an organisms growing at the expense of one substrate also can transform a different substrate that is not associated with that organism's energy production, carbon assimilation, or other growth processes.

Since co-metabolism generally leads to a slow degradation of the substrate, attention has been given to enhancing its rate. The addition of a number of organic compounds into the contaminated zone may promote the rate of co metabolism, but the responses to such additions are not predictable. Addition of mineralizable compounds that are structurally analogous to the compound whose co-metabolism is desired is known as analog enrichment (Alexander, M., 1994). The microorganisms that grow on the mineralizable compound contain enzymes transforming the analogous molecule by co-metabolism.

Microorganisms are capable of catalyzing a variety of reactions: *dechlorination* (the chlorinated compound becomes an electron acceptor; in this process, a chlorine atom is removed and is replaced with a hydrogen atom), *hydrolysis* (frequently conducted outside the microbial cell by exo-enzymes). Hydrolysis is simply a cleavage of an organic molecule with the addition of water, (cleaving of a carbon-carbon bond is another important). An organic compound is split or a terminal carbon is cleaved off an organic chain, *cleavage* (cleaving of a carbon-carbon bond is another important reaction). An organic compound is split or a terminal carbon is cleaved off an organic chain, *oxidation* (breakdown of organic compounds using an electrophilic form of oxygen), *reduction* (breakdown of organic compounds by a nucleophilic form of hydrogen or by direct electron delivery), *dehydrogenation* (an oxidation-reduction reaction that results in the loss of two electrons and two protons, resulting in the loss of two hydrogen atoms), *Dehydro-halogenation* (an oxidation-reduction reaction that results in the loss of two electrons and two protons, resulting in the loss of two hydrogen atoms), and *substitution* (these reactions involve replacing one atom with another) (Suthersan, 1999).

BIOREMEDIATION POTENTIAL OF WHITE-ROT FUNGI

White Rot Fungi (WRF) comprises the only group of organisms that can degrade wood completely possess unique biodegradation potential (Aust, 1990; Bumpus *et al.*, 1985;

Hammel, 1992; Field *et al.*, 1993). To develop efficient and environmentally suitable utilization of lignocellulosics, biotechnological approaches have received great attention. The degradation mechanism of lignin by microorganisms and enzymes should be elucidated to gain background knowledge in biochemical processing of lignocellulosics. Since a mini review article (Pointing, 2001) on a similar subject was published in 2001, great progress has been made in biodegradation of pesticides, munitions waste, organochlorines, polychlorinated biphenyl, PAHs, bleach plant effluent, synthetic polymers, synthetic dyes. Biobleaching of chemical pulps, improving mechanical pulps and wastewater treatments by lignin-degrading fungi and their enzymes has been accepted as potential application (Higuchi, 1987).

Lignin is a very complex, heterogeneous aromatic polymer containing various non-repeating phenyl propanoid units linked by various biochemically stable carbon-carbon and ether linkage between monomeric phenyl propane units (Sakakibara, 1983).

Possibility of using white rot fungi for bioremediation strategies in recent years, has initiated extensive research effort in government institutions, academic and industries. The interest in area arises from the capability of white rot fungi to degrade an enormously diverse range of very persistent or noxious environmental pollutants. This ability makes the use of white rot fungi apart from many of the existing methods of bioremediation due to the stereo irregularity of lignin makes it very resistant to attack by enzymes. In addition, it is impossible for lignin to be absorbed and degraded by intracellular enzymes.

Perhaps the easiest way to understand the nonspecific ability of these fungi to degrade pollutants is to consider their ecological "niche." White rot fungi are those organisms that are able to degrade lignin, the structural polymer found in woody plants. The white-rot fungi are a physiological rather than taxonomic grouping. In nature, fungi do much of the dirty work. They are efficient at degrading the major plant polymers, cellulose and lignin, but they also decompose a huge array of other organic molecules including waxes, rubber, feathers, insect cuticles, and animal flesh.

Excellent reviews and research papers have appeared in the last three decades describing the mechanism and process of potential of white rot fungi (WRF) to degrade lignin and application of ligninolytic enzyme systems in cleaning of environmental pollutants

Aust and Benson, 1993; Bassham, 1975; Boominathan and Reddy, 1992; Buswell and Odier, 1987; Buswell *et al.*, 1984; Collins and Dobson, 1995; Cripps *et al.*, 1990; de Jong *et al.* 1994; Eaton and Hale, 1993; Eggert *et al.*, 1996; Farrell *et al.*, 1989; Field *et al.*, 1993; Glenn and Gold, 1983; Glenn *et al.*, 1986; Gosczcynski *et al.*, 1994; Gregory, 1993; Harvey *et al.*, 1985a, 1986b; Hatakka, 1994; Hatakka and Uusi-Rauva, 1983; Heinfling *et al.*, 1998; Higuchi, 1986; Jeffries, 1990; Kirk and Chang, 1981; Kirk and Farrell, 1987; Kirk, 1981a; Kirk, 1981b; Leonowicz *et al.*, 1999; Michaels and Lewis, 1985; Ollikka *et al.*, 1993; Orth and Tien, 1995; Palmer *et al.*, 1986; Pasczcynski *et al.*, 1991; Pasti and Crawford, 1991; Pasti-Grigsby *et al.*, 1992; Pasczçynski *et al.*, 1992; Pointing and Vrijmoed, 2000; Pointing *et al.*, 2000b; Pointing, 2001; Reddy and D'Souza, 1994; Reddy, 1995; Rodriguez *et al.*, 1999; Schoemaker *et al.*, 1985; Spadaro *et al.*,1992; Thurston, 1994; Tien and Kirk, 1983, 1984; Vyas and Molitoris, 1995; Wariishi *et al.*, 1992.

AIMS AND OBJECTIVES

Considerable attention has been paid to the color removal of effluents generated from dyeing houses during the processes of dyeing, printing, and washing of textile fabrics. Since, the physico-chemical treatments methods have not fulfilled the process of decolorization of dye residues in the effluents, more attention was paid towards the utilization of biological processes to remove the color.

The present study aimed to explore microbial populations, having potential to decolorize and degrade the dyes, from the sources such as textile effluent treatment plant. It was also aimed to find out the physiological conditions necessary, for the organisms isolated, under which they decolorize the dyes.

In *this study* following *approaches* were considered:

1. Identification of related sites for sample collection.
2. To isolate and screen indigenous microorganisms from samples to select organisms capable of decolorization of selected dyes under laboratory conditions.
3. To characterize the isolated organisms based on phenotypic characters and antibiotic sensitivity assay.
4. To establish rates of decolorization under different cultural conditions.
5. To study, one of highly potential for its potential to decolorize dye, when subjected to various nutritional conditions.
6. To obtain qualitative information of degraded products by TLC, and UV-Visible spectroscopy.
7. To demonstrate the fate of dye precursors-aromatic amines when subjected to the organism in the medium by chromatographic profile study.
8. To find out the potential of the isolates to degrade PAHs.
9. To study the capability of white rot fungi on decolorization of various textile dyes.

2

MATERIALS AND METHODS

SAMPLING

Sampling Site : Common Effluent Treatment Plant

To search for microorganisms having ability to decolorize dyes, the sampling site selected was common effluent treatment plant of Jetpur textile industries. The plant is located outside (3-4 kilometers) the Jetpur town, which receive wastewater from dyeing and printing houses at and around Jetpur–Navagadh town complex.

Sample Collection

Five samples from different stages of a common textile effluent treatment plant (where the effluents from all textile industries unruffled) were collected in autoclaved BOD bottlesand kept at 4°C immediately after collection and tested within 24 hours and labeled as shown below.

TABLE 2.1 : TABLE SHOWING STAGES, SITE OF COLLECTION AND COLOR OF EFFLUENTS

Stage No.	Sample	Collection Site	Color of Effluents
Stage-1	UT-1	Raw Collection Tank	Reddish Brown
Stage-2	UT-2	Filtered Tank	Brown
Stage-3	SF-	Activated Sludge Tank	Brown
Stage-4	S	Activated Sludge Tank	Greenish
Stage-5	T	Treated Collection Pond	Greenish

PHYSICO-CHEMICAL CHARACTERISTICS OF THE SAMPLES

Some common characteristics of effluents like pH, COD, BOD etc. were analyzed as per APHA (Andrew *et al.*, 1995).

MICROBIOLOGICAL ANALYSIS OF THE SAMPLES

Diversity of Microbial Population of CETP

A Complete Medium Broth (CMB) consisting of (gl^{-1}): KH_2PO_4 10; yeast extract 5; peptone 5; glucose 2; $MgSO_4 .7H_2O$ 0.02, having pH, 7.0, was prepared in 500 ml of Erlen Meyer (EM) flask containing 200 ml of medium and a specific volume of effluents was added into the medium and incubated on shaker at 100 rpm for one week. During incubation period, every after two days, a loopful medium was streaked onto nutrient agar plates and then incubated at 30°C for 24 to 72 hours. Colonies from plates were observed after incubation were subcultured and then pure culture was obtained. The isolates of microbial population obtained were grouped on the basis of cell morphology, Gram's reaction and cultural characteristics. Results were expressed as percentage distribution of the organisms isolated from samples of different stages of CETP.

Enrichment and Isolation of Dye Decolorizing Microbial Populations

Samples collected in the sterile BOD bottles from the CETP were analyze for the existing microbial population, employing the Complete Medium Broth (CMB) consisting of (gl^{-1}): KH_2PO_4 10; yeast extract 5; peptone 5; glucose 2; $MgSO_4$, $7H_2O$ 0.02; pH 7.0, in 500 ml of EM flask containing 100 ml of the medium. Separately autoclaved solution of a mixture of 20 dyes (Table 2.2) was added in the medium to make the final concentration 0.02% of each dye. Each flask was inoculated with 1 ml of the samples and incubated for extended period under static and shaking condition (100 rpm), at 30°C. The flasks were observed everyday for growth and disappearance of color. 2 ml (20 mg) of the dye mixture solution was added in the same flasks in which color had disappeared, and incubation continued. A loopful of inoculum was transferred in the same fresh medium containing dyes mixture solution and further incubated. Thus, the transfers were repeated three times.

The flasks in which growth and decolorization observed were selected and a loopful of sample was transferred on the Complete Medium Agar (CMA) plates containing the above-mentioned ingredients, including solution of the dye mixtures, incubated at 30°C for 24 to 96 hours. Colony developed on the plates had been studied for colony characteristics. The selected colonies were transferred on the fresh medium plates for further study. All isolated colonies from at least five plates containing 5 to 25 colonies with different morphology were transferred on CMA plates to obtain pure culture. Pure cultures of each bacterial isolate were obtained after three times transferred on agar plates and then inoculated on agar slants for maintenance.

The screening of the bacterial strains for dye decolorization was carried in 250 ml EM flask containing 150 ml of CMB medium and incubated 30°C, under static condition. For one month mixture of all dyes (10 mg/ml) listed (Table 2.2) was added regularly in the flasks in which color was disappearing.

From the primary screening study, fifty-five bacterial isolates were observed and maintained on agar slants. All the fifty-five isolates were coded as kot-01 to kot-55.

Grading of the Bacterial Isolates Based on Decolorization Capabilities

These 30 isolates were further investigated for their ability to decolorize the twenty dyes and graded as poor decolorizer, moderate decolorizer and highly potential decolorizers. The results are listed in (Table. 3.2 and 3.3).

Preparation of Dye Solution

Solution of all the 20 dyes was prepared in double distilled water (50 mgl^{-1}). The UV-Visible spectra of each dye solution were taken spectrophotometrically to determine the wavelength at which the dye molecules gave the highest absorbance. From these results, the λ_{max} of each dye was selected (Table. 2.2).

TABLE 2.2 : LIST OF TEXTILE DYES

Sr. No.	Dye Code	Name of Dyes	λ_{max} nm	Generic Name	Structure
01	Jd-00	Kemifix Red-F6B	530	Cl-Reactive Red-250	Available
02	Jd-01	RS Red H5BL	512	Not Available	Not Available
03	Jd-02	RS Violet H5RL	560	Cl: Reactive violet-5	Not Available
04	Jd-03	RS Navy Blue H2GL	596	Cl: Reactive Blue-203	Not Available
05	Jd-04	RS Brown HGRL	482	Cl: Reactive Brown-18	Not Available
06	Jd-05	Red H8B	527	Cl: Reactive Red-31	Not Available
07	Jd-06	Red M5B	542	Cl: Reactive Red-2	Not Available
08	Jd-07	Purple H3R	565	Not Available	Not Available
09	Jd-08	Orange 3R	500	Cl: Reactive Orange-16	Not Available
10	Jd-10	Reactive Blue-81	664	Cl: Reactive Blue-81	Available
11	Jd-12	Acid Red-249	543	Cl: Reactive Red-249	Available
12	Ad-13	Pc-Black-HN	583	Not Available	Not Available
13	Ad-14	Red-HB	520	Not Available	Not Available
14	Ad-15	T-Blue-G	670	Cl: Reactive Blue-21	Not Available
15	Ad-16	Yellow-FGM	425	Not Available	Not Available
16	Ad-17	Red 6 Bx	545	Not Available	Not Available
17	Ad-19	Reactive Yellow-4	440	Cl: Reactive Yellow-4	Available
18	Kd-22	Red Brown HoR	551	Not Available	Not Available
19	Kd-23	Golden-HR	422	Cl: Reactive Orange-12	Not Available
20	Kd-24	Magenta-HB	564	Not Available	Not Available

Identification of the Bacterial Isolates

Attempts were made to identify the confirmed decolorizing bacterial isolates by studying following phenotypic characteristics as given below as per Bergey's manual (1994).

- ***Growth characteristics***
  - Growth on Nutrient Agar Plate (Colony characteristics)
  - Growth on Agar Slant
  - Growth in Nutrient Broth
- Gram's reaction and cell morphology
- Biochemical tests

GROWTH CHARACTERISTICS

Growth on Agar Plates (Colony Characteristics)

Young cultures were prepared of all isolates in 250 ml of EM flask containing 100 ml of CM broth. Simultaneously, CMA plates were prepared and kept for 24 hours to check whether any contamination is there or not. A loopful culture was streaked out on CMA to obtained isolated colonies and incubated at 30°C for 24 hours. In this study four-flame method was employed for streaking. Next day, isolated colonies were marked and colony characteristics were noted down (Table 3.4).

Growth on Nutrient Agar Slant

Isolated colonies were selected and from a single colony, six hours young culture was prepared and a loopful culture was streaked on nutrient agar slant. All the slants were incubated at 30°C for 24 hours and then growth characteristics on nutrient agar slants were examined (Table 3.5).

Growth in Nutrient Broth

A loopful culture of all isolates from slants was inoculated in nutrient broth. All flasks were incubated at 30°C for 24 hours and the growth characteristics like uniform growth, pellet growth, sediment growth etc. were observed (Table 3.6).

Gram's Reaction and Cell Morphology

Gram's reaction of all the isolates was performed and on the basis of microscopic observation the cell morphology study was carried out.

Biochemical Tests

Media for all biochemical tests as listed (Table 3.7, 3.8, 3.9, 3.10, 3.11, and 3.12) were prepared and as mentioned above, young culture were prepared and a loopful of culture were inoculated, then all media were incubated at 30°C for 24 hours and results were examined.

Identification of Bacterial Isolates

On the basis of all the tests performed as above, mentioned tentatively identification and categorization of the bacterial isolates were made and results were tabulated (Table 3.13).

ANTIBIOTIC SENSITIVITY TESTS

Materials

Antibiotic discs (HiMedia Laboratories Ltd., Octadiscs, B. No. 0-2167, U.B.B-2003, previously stored at 0 to 8°C) having fixed concentration of each antibiotic were used.

In this study antibiotics (mentioned in following table) were tested for all thirty isolates under following standard conditions.

Methods

The antibiotic sensitivity of bacteria can be assessed in a variety of ways according to individual preference, the constraints of cost, the nature of the bacterium, the number of strains requiring investigation and the degree of accuracy required. Traditional methods fall into one of three

S. No.	Name of Antibiotic	Abbreviation
1.	Cephalothin	Ch
2.	Clindamycin	Cd
3.	Co-Trimoxazole,	Co
4.	Erythromycin	E
5.	Gentamycin,	G
6.	Ofloxacin,	Of
7.	Penicillin-G	P
8.	Vancomycin	Va
9.	9Augmentin	Au
10.	Co-Cethotaxime	Ce
11.	Tobramycin	Tb
12.	Gentamycin,	G
13.	Piperacillin	Pc
14.	Ceftazidime-G	Ca
15.	Coistin	Ci
16.	Amikacin	Ak
17.	Ampicillin	A
18.	Cephoxitin	Ca
19.	Ceftazidine	Ci
20.	Ceftriaxone	C
21.	Chloramphenicol	G
22.	Piperacillin	Pc
23.	Tetracycline	T
24.	Carbenicillin	Cb
25.	Cephoxitin	Cn
26.	Chloramphenicol	C
27.	Erythromycin	E
28.	Metronidazole	Mt

main categories, agar diffusion tests, in which the antibiotic is allowed to diffuse from a point source, commonly in the form of an impregnated filter paper disc, into an agar medium that has been seeded with the test organism, broth dilution tests, in which serial (usually two fold) dilutions of antibiotic in a suitable fluid medium are inoculated with the test organism. The highest dilution of the antibiotic to prevent the development of visible growth after overnight incubation is the Minimum Inhibitory Concentration (MIC) and agar incorporation tests, which are essentially similar to broth dilution tests except that the antibiotic dilutions are incorporated in an agar medium in a series of Petri dishes. These are spot-inoculated with a number of test organisms, usually by means of a semi-automatic inoculating device.

Agar Diffusion Test

Most diagnostic microbiology laboratories test antibiotic sensitivity of bacteria by some form of agar diffusion test in which the organisms under investigation are exposed to a diffusion gradient of antibiotic provided by an impregnated disc of filter paper. When the bacterial population reaches a certain critical concentration, no further inhibition of growth can be achieved and the edge of an inhibition zone is formed. Many antibiotics can be tested on one culture plate. A zone size of 6 mm indicates no zone of inhibition, since this is the diameter of the antibiotic-containing disc

Preparation of Plates

Nutrient agar medium was prepared, autoclaved at 121°C, 15 lbs pressure for 20 minutes and used for sensitivity testing. Before inoculation plates, were allowed to dry with the lids, so that there are no droplets of moisture on the agar surface.

Preparation of Inoculums

Six-hour young culture of all isolates under testing were prepared in nutrient broth at 30°C temperature. Culture density was maintained constant in all the antibiotic assays throughout the experiments.

Inoculation and Incubation

Nutrient agar with uniform thickness was prepared adding 15 ml of melted nutrient agar, it was allowed to solidify and then, 10 ml of top nutrient agar with 0.1 ml of six-hour young culture of respective isolates was added. The melted agar tubes then were poured and allowed to solidify. Antibiotic discs were applied under aseptic condition with forceps to ensure even contact with the medium. At a time eight discs of antibiotics (previously stored at 4°C in sealed container with a desiccant) were applied in single petriplate, before the containers were opened, expiry date was confirmed. All plates after inoculation were incubated overnight at 35°C to 37°C temperature and results were tabulated.

SELECTION OF BACTERIAL STRAINS BASED ON THEIR EFFICIENCY TO DECOLORIZE DYES

Among the fifty-five bacterial cultures isolated, the isolates showing higher potential to decolorize the twenty selected dyes were further examine under the same cultural conditions describe earlier. Out of the fifty-five cultures, when subjected for their decolorization potential, the actual number reduced to thirty bacterial isolates.

These thirty bacterial isolates were investigated for their identification.

Since it was not feasible to conduct experiments on decolorization of 20 dyes with the 30 bacterial strains, it was decided to further screen the highly potential organisms using only some selected dyes.

Thorough examination of the data recorded on the decolorization of 20 dyes with 30 isolates, it was found that, the three dyes (Kemifix Red F6B, RS Red H5BL, and RS Brown HGRL) were decolorized by the 13 strains out of all the 30 strains.

Therefore, various experiments were conducted with combinations of these 13 bacterial strains and the 3 dyes as follow:

- The 13 designated bacterial strains (Kot-01 to Kot-10, Pseu-I, Pseu-II, and Pseu-III) were examined for their percentage decolorization and extent (time required) of decolorization of each dye.
- Activated cultures of each strain were prepared by inoculating culture flasks containing CMB medium (100 ml in 250 ml EM flask), incubated at 30°C under shaking condition.
- After incubation of 12-16 hours the medium was harvested and centrifuged. The pellets were resuspended in saline solutions to make 1.0 OD of the suspension. All the steps were carried out under aseptic conditions.
- 2 ml of the cell suspension was added to CMB medium, incubated at 30°C, under shaking condition for exactly 6 hours.
- 2 ml of activated culture (approx. 1.0 OD) of each bacterial strain was transferred to the EM flasks containing CMB medium (100 ml) and each dyes solution having final concentration of 0.02%.
- The inoculated flasks were incubated at 30°C under static and shaking conditions.
- The observations were recorded by monitoring the decolorization of dyes by spectrophotometrically at every 24, 48, and 72 hours. The calculated results were expressed as percentage decolorization vs. time and presented graphically.

BATCH ASSAYS OF DECOLORIZATION OF DYES BY *PSEUDOMONAS Sp.* STRAIN-II

In the present experiment, *Pseudomonas* sp. Strain-II was used to determine quantitatively dye decolorization and its response in presence of various combinations of nutritional parameters under static condition. The seed culture was prepared by cultivating the organisms in the CMB medium as describe previously.

Effects of Nutrient Constituents in Different Combinations

Various nutrient constituents in the medium were examined to study their effects on decolorization of dyes. Following combinations of ingredients were tested using dye RS Red H5BL (λ_{max} = 512 nm).

1. Glucose + KH_2PO_4 + $MgSO_4$,$7H_2O$ + Dye = [MGD]

2. Yeast extract + KH_2PO_4 + $MgSO_4$,$7H_2O$ + Dye = [MYED]

3. Peptone + KH_2PO_4 + $MgSO_4$,$7H_2O$ + Dye = [MPD]

4. Benzoic acid + KH_2PO_4 + $MgSO_4$,$7H_2O$ + Dye = [MBAD]

5. Glucose + peptone + KH_2PO_4 + $MgSO_4$,$7H_2O$ + Dye = [G1]

6. Glucose + yeast extract + KH_2PO_4 + $MgSO_4$,$7H_2O$ + Dye = [G2]

7. Yeast extract + peptone + KH_2PO_4 + $MgSO_4$,$7H_2O$ + Dye = [G3]

8. BA + peptone + KH_2PO_4 + $MgSO_4$,$7H_2O$ + Dye = [P1]

9. BA + yeast extract + KH_2PO_4 + $MgSO_4$,$7H_2O$ + Dye = [P2]

In the above media the ingredient contents was in (g/l^{-1}): glucose 2; yeast extract 5; peptone 5; BA 5; KH_2PO_4 10; and $MgSO_4$ $7H_2H$ 0.02; dye 0.2; pH 7.0.

Activated cultures of organisms were prepared as described in preparation of dye solution. 2 ml of the inoculum (approx. 1.0 OD) was added in the respective media (01 to 09 combinations as given above) and incubated at 30°C under static condition for 60 hours. All the readings were taken in triplicates. At regular interval of four hours, 2 ml of assay medium was pipetted, transferred in eppendorf tubes and centrifuged, absorbance was measured at λ_{max}, 512 nm spectrophotometrically. The measurements were continued upto 60 hours. The sediments were suspended in normal saline solution and again centrifuged. Supernatant was discarded, sediments in the tubes was oven dried overnight at 80°C and dry weight was measured (mg/ml) to estimate biomass. The biomass was plotted against reduction in color with respect to time.

DECOLORIZATION AND DEGRADATION ASSAY OF DYES BY *PSEUDOMONAS Sp.* II

Documentation of Biodecolorization in Assay Mixture

10 ml of CMB medium was distributed to each test tube and autoclaved. Each dye solution was autoclaved separately and added in the medium to achieve final concentration of 0.02%. Finally all the tubes, except controls, were inoculated by adding 2 ml of young culture (6 hrs, having 1.0 OD) having approximately 1×10^8 colony forming units per milliliter (CFU/ml) of the bacterium *Pseudomonas* sp II. Decolorization of the dyes was visually observed within 24 hours of incubation.

UV-Visible Spectrophotometric and Chromatographic Analysis of Assay Mixture

Spectrophotometric Analysis

2 ml of assay mixtures of seven dyes (Kemifix Red F6B, RS Red H5BL, RS Violet H5RL, RS Navy Blue H2GL, RS Brown HGRL, Red H8B, and Red M5B) in the incubated test tubes used in the previous experiments were harvested and centrifuged at 10.000 RPM for 15 minutes. Supernatant was transferred in cuvettee and employed to measure spectra by scanning (190 to 1100 nm) using UV-Visible spectrophotometer (Elico-164 UV-Visible spectrophotometer).

TLC Analysis of the Samples

- TLC Plates Silica Gel $60F_{254,}$ having layer thickness of 0.25 mm × 5 cm × 10, Merck K. GaA- 64271, Darmstadt, Germany, coated on aluminium foil was used for analysis.

- Several polar and non-polar solvents were tested to standardize better solvent systems to determine Rf values to separate the components in the medium before and after incubation of assay mixtures with 20 dyes. Polar solvents like methanol, acetone and ethyl acetate and non-polar solvents like benzene, hexane, chloroform and dichloromethane in various combinations were employed to standardize adequate clear separation of the dyes and degraded products.
- 100 microlitre of the samples were diluted to 1000 microlitre and used for loading on the TLC plates.
- Before loading the TLC plates were activated for one hour samples were loaded on TLC plates and labeled as 1 (dye only), 2 (medium only) 3 (medium+ organisms), and 4 (medium + dye + organisms treated), and allowed to run in two standardized solvent systems (Methanol: Chloroform and, Methanol: Ethyl acetate in proportion of 9:1 for both the solvent systems) previously saturated with the same.
- Dried TLC plates were examined in UV chamber under short (254 nm) and far (365 nm) to examine the fractions of dye and/or medium.
- Rf value of the 20 dyes were calculated and tabulated (Table 3.15).

DEGRADATION OF AROMATIC AMINES

The aim of this study was to investigate which aromatic amines (basic chemical moiety of azo dyes) could be readily metabolized the bacterial strain under aerobic conditions. In this study the biodegradation of a mixture of two aromatic amines, Benzidine and 4-chloroaniline and an aromatic amine, 2-naphthylamine was examined in batch culture experiment using *Pseudomonas* sp. II.

Table 2.3 shows the list of some selected amines used in the synthesis of textile dyes.

TABLE 2.3 : LIST OF SOME AROMATIC AMINES USED IN SYNTHETIC TEXTILES

Sr No.	C.A.S	Name of Amines	Retention time (min.)	Chemical Structures
1	[95-53-4]	O-Toluidine*	6.38	
2	[97-56-3]	O-Aminoazotoluene*		
3.	[95-80-7]	2.4-Diaminotoluene**	11.4	
4	[137-17-7]	2,4,5-Trimethylaniline	9.62	
5	[99-55-8]	2-Amino-4-nitrotoluene**	-	
6	[60-09-3]	4-Phenylazoaniline	?	
7	[90-04-4]	2-Methoxyaniline	7.9	
8	[91-59-8]	2-Naphylamine	13.08	
9	[91-94-1]	3.3'-Dichlorobenzidine	29.75	
10	[92-67-1]	4-Amonobiphenyl	15.36	
11	[92-87-5]	Benzidine	21.19	

Sr No.	C.A.S.	Name of Amines	Retention time (min.)	Chemical Structures
12	[95-69-2]	4-Chloro-o-toluidine	9.72	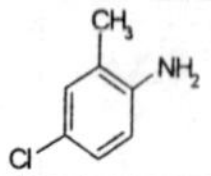
13	[106-47-8]	p-chloroaniline	8.36	
14	[615-05-6]	2,4-Diaminoanisole	7.94	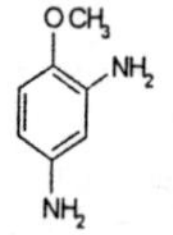
15	[101-77-9]	4,4'-Diaminodiphenyl methane	21.31/23.36	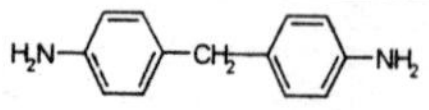
16	[119-90-4]	3,3'-Dimethoxybenzidine	25.74	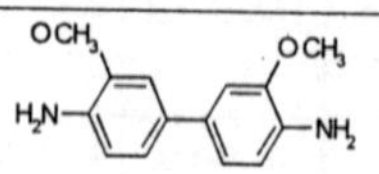
17	[119-93-7]	3,3'-Dimethylebenzidine	25.74	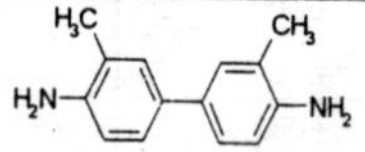
18	[838-88-0]	3,3'-Dimethyle-4-4'-diaminophenylemethane	25.16/27.97	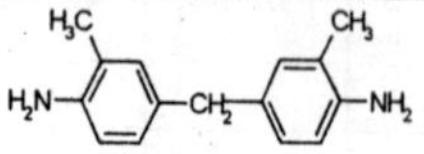
19	[120-71-8]	p-Cresidine	9.45	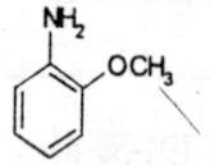
20	[101-14-4]	4,4'-Methylene-bis-(2-chloroaniline)	30.01	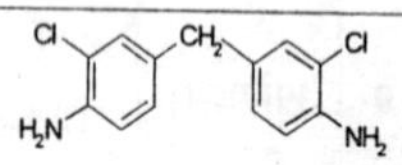
21	[101-80-4]	4,4'-Oxydianiline	20.97	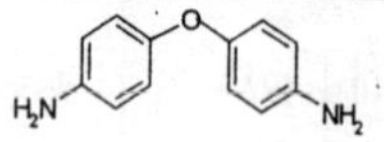
22	[139-65-1]	4,4'-Thiodianiline	27.19	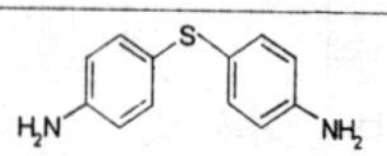

The CMB medium was used in which the aromatic amines were added from the autoclaved and neutralized (pH 7.0) stock solution. 250 ml of EM flasks containing 100 ml of medium with 500 ppm concentration of each aromatic amine (benzidine and 4-chloroaniline) were inoculated with 6 hours young culture of *Pseudomonas* sp.-II and incubated at 30°C on shaker at 100 rpm, for 24 hour and then analysis was carried out after centrifugation, by HPLC and GC.

Sample Preparation

The supernatant was used with diethyl ether (liquid-liquid extraction) and allowed to evaporate till dryness and then it was solubilized in methanol and was used for HPLC and GC analysis.

Analysis

The methanol soluble samples (filtered through 0.2 micrometer filter) were analyzed by HPLC [Agilent 1100 HPLC, Column: Zorbax Eclipse-XLB, having detector: Diode Array Detector (DAD)]. During the chromatographic run, the following conditions were maintained.

Mobile Phase

A: Methanol, **B:** (0.5758 ammonium dihydrogen phosphates +0.78 disodium hydrogen phosphate) in 1000 ml of distilled water (HPLC grade).

Mobile Flow Program: Gradient

Time	Flow	A	B
0.0	1 ml	15%	85%
45	1 ml	80%	20%

For gas chromatography (Fison's MD 800) the following set of conditions were used.

Oven Programme

GC (Inlet) : Column DB-5 ms

60°C (1 min.) : 200 @ 10°C/min;—230°C @ 2°C/min.; 250°C @ 15°C/min. (5 min.)

MS (Detector)

Solvent delay : 05 mins.

Full Scan : 50.00 to 350.00 amu.

Ionization : EI+

RING FISSION REACTION INDICATING DEGRADATION OF DYES

The *ortho* or *meta* cleavage of two central intermediates in the metabolism of aromatic compounds (catechol and protocatechuate) can be tested by the method originally suggested by Dr. K.Hosokawa (Stanier *et al.*, 1966).

Procedure

1% stock solution of dyes Kemifix Red F6B, RS Red H5BL, RS Violet H5RL, RS Navy Blue H2GL, and RS Brown HGRL was prepared and autoclaved separately. CMB medium was prepared and autoclaved. Before inoculation, the autoclaved dyes were added to make 0.01% final concentration in the medium and were inoculated with 6-hour young culture of *Pseudomonas* sp.-II. The culture flasks were incubated at 30°C in static condition. The entire test was carried out in triplicate alongwith control without inoculation of dyes but with organisms. 10 ml of suspension was harvested after incubation of 72 hours from each flask, centrifuged at 10,000 rpm for 10 minutes at 4°C.

The Rothera Reaction

➢ Meta ***Cleavage of Ring***

The harvested pellets were suspended in 2 ml of 0.2 M Tris-buffer (pH 8.0) and 0.2 ml of toluene added to solubilize the cell membrane. 0.2 ml of 1.0 M catechol or protocatechuate was added and thoroughly mixed. The assay mixture was confirmed as positive based on formation of yellow color after few minutes, indicating *meta* cleavage activity.

➢ Ortho *Cleavage*

1 gm of $(NH_4)_2SO_4$ was added in 2.5 ml of the cell suspension. The assay mixture was incubated for 1 hour at 30ºC. After incubation, pH of the mixture was adjusted ~10.0 by adding 0.5 ml ammonium solution (5 N) and a drop of 1% sodium-nitropruside. Positive activity of *ortho* cleavage in the assay mixture was confirmed by observing formation of deep purple color.

Ring Fission Reaction Indicating Degradation of PAHs

As mentioned earlier (Ring Fission Reaction Indicating Degradation of Dyes) the ring fission mechanism was performed for PAHs also to test whether bacterial sp., which is under experiment, is able to degrade PAHs or not.

Procedure

- Stock solution of PAHs (Table 2.4) was prepared and autoclaved separately. Mineral Salt Medium (MSM) containing (g/l): KH_2PO_4 0.7; K_2HPO_4 0.7; $MgSO_4$, $7H_2O$ 0.7; NH_4NO_3 1.0; NaCl 0.005; $FeSO_4.7H_2O$ 0.002; $ZnSO_4.7H_2O$ 0.002; $MnSO_4.7H_2O$ 0.001; and selected PAHs. pH of the medium was set 6.4. In one set 0.2% glucose was added in one set glucose was added as co-substrate and in another set glucose was not added and all flask were incubated at 100 rpm on shaker for three days.

- After incubation period first of all growth response (in terms of +, ++, +++, ++++) was noted down and then the sample were tested for ring cleavage test. The ring cleavage test was performed as discussed earlier as follow and results were tabulated.

WHITE ROT FUNGI AND DYES

Chemicals

Chemicals used in our study were mostly of the highest purity available. Dyes (listed previously) were obtained from Jetpur Textile Industry's Association.

Microorganisms

Lignin degrading white rot fungi (WRF) *Coriolopsis polysona* (CCBAS 740) (Sasek *et al.*, 1993), and a newly isolated white rot fungus sp Rc were maintained on malt extract agar slants. Transfer were made on Fortified Malt Extract Agar Medium (FMEM) from these stock cultures and cultivated at room temperature.

Culture Media and Cultivation of WRF

FMEM used for the cultivation of WRF as petriplate culture. (Tien and Kirk, 1988) consisted of the following ingredients (gl^{-1}): glucose, 10; malt extract, 10; yeast extract, 2; peptone, 2; KH_2PO_4, 2; $MgSO_4.7H_2O$, 1; aspargine, 1; thiamine—HCl, 1 mg; agar, 25; pH 4.2.

Production of Ligninolytic Enzyme by WRF by Solid-State Fermentation (SSF)

Chopped wheat straw (2 to 3 cm.) was washed thoroughly under running tap water and oven dried. 5 gm of wheat straw was taken in 250 ml EM flasks and 20 ml of distilled water was added. These flasks were sterilized at 121°C, for 20 minutes. This cycle was repeated three times.

Inoculum and SSF

Ten agar discs punched from FMEM plate cultures of WRF were used to inoculate single flask. These flasks were incubated at room temperature and harvested when the content of flask was covered with fungal mycelial growth (generally after 14 days) for enzyme extraction.

Extraction and Preparation of Crude Extracellular Enzyme

A set of twenty flasks was harvested and processed immediately for the preparation of extracellular enzyme extract as described previously (Vyas and Molitoris, 1995). To the culture flask, 100 ml phosphate buffer (0.1 M, pH 6.5) containing 0.1 M, NaCl was added. The contents were gently beaten to break the clumps and incubated on rotary shaker for one hour. The extract was then filtered and centrifuged at 5000 rpm for 15 minutes at 4°C temperature. Enzyme extract was then subjected to ammonium sulphate precipitation (80% saturation) and the precipitates were separated upon centrifugation at 5000 rpm, for 15 minutes at 4°C and were dissolved in minimum amount of phosphate buffer (0.1 M, pH 6.5). The required quantity of enzyme extract was thawed and clear supernatants obtained upon spinning, referred as Concentrated Extracellular Enzyme Preparation (CEEP), was then used for various biochemical studies.

Dye Decolorization Assay

Reaction mixture (RM) for MnP consisted of 1000 μl, Succinate Lactate Buffer (SLB) (0.1 M, pH 4.0 and 4.5), 50-200 μl dye (0.1%), 10 μl $MnSO_4$ (20 mM), 70 μl H_2O_2 (10 mM) and enzyme sample.

Reaction mixture (RM) for MIP consisted of 1000 μl. Tartrate Buffer (TB) (0.1 M, pH 2.5, 3.0, and 3.5), 50-200 μl dyes (0.1%), 70 μl H_2O_2 (10 mM) and enzyme sample.

Reaction mixture for Laccase consists of 1000 μl TB (0.1 M, pH 2.5, 3.0, and 3.5), 50-200 μl dyes (0.1%), and enzyme sample.

Control consisted of identical reaction mixture except that enzyme added was either heat killed (incubating ES in boiling water bath for 3-5 min) or native enzyme added at the end of incubation. CEEP was added prior to the initiation of reaction with H_2O_2 addition. Reaction was monitored continuously at fixed wavelength (λ_{max} of the test dyes) using UV-Visible spectrophotometer (Systronic-119). Reaction mixture in reference cuvette contained all the components of the experimental except dye.

3

RESULTS

DESCRIPTION OF THE SITE

The dyeing and printing units are the major industries in and around the Jetpur town (Dist. Rajkot, Gujarat State, India). The dyeing and printing had been a cottage industry since the beginning of the 19th century. However, the industry in its present state developed since 1947, which gradually replaced manual to mechanical methods. The industries reached to peak during 1980s.

The reported data revealed that the dyeing and printing houses at and around Jetpur are highly water consuming and the water consumed becomes highly contaminated with dyeing-printing-finishing chemicals, during various processes. The contaminated wastewater is discharged in the environment. The Jetpur industrialists have financially supported a common collection, conveyance and treatment system for the wastewater to alleviate pollution problems. However, there appears to be still considerable room for improvement (Anonymous, 1995).

The data on water consumption and discharges in dyeing/printing units at and around Jetpur is presented in Fig. 3.1. The industry is by far the greatest consumer of water and especially the Ghat (70% of the total consumption), and does so almost exclusively from groundwater resources.

In 1992, the treatment plant was commissioned. The total wastewater flow was measured to be around, 16000 m^3/d, and the plant was designed for 20,000 m^3/d. Several conventional

FIG. 3.1 : SHOWING VALUES OF WATER CONSUMPTION AND DISCHARGE OF EFFLUENTS

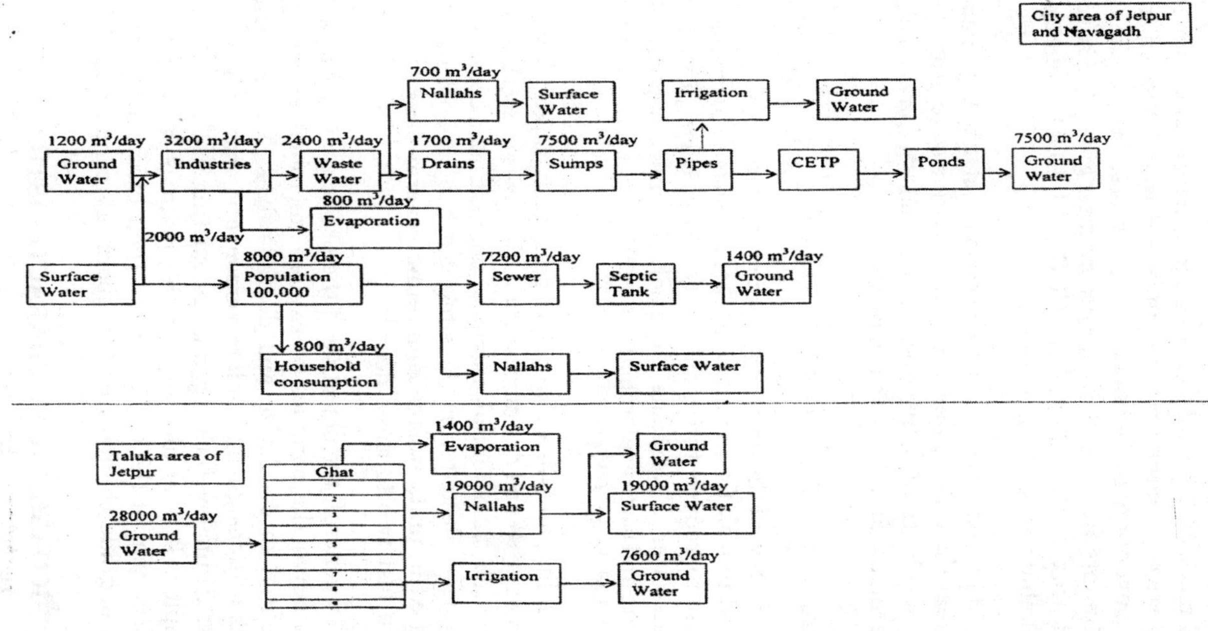

and non-conventional treatment techniques have been develop and tried at industrial scale in recent times. Viewed in terms of conventional wastewater parameters, out of the several pollutants discharged in the effluent guar gum, starch, softener, and wax contribute to the COD and BOD. Remazol dyes contribute to COD and color.

At present, the treatment system seemed to be not adequate to treat the wastewater, especially, the dye color problem, since the whole treatment process is based on physico-chemical treatment, and extended aeration.

GOALS OF THE PRESENT STUDY

In view of the above discussion, the present study has been aimed to look for the biotic agents of native environment that assist to reduce the high levels of dye/color present in the wastewater. To execute the goal, the main objective was to isolate and screen microbial populations occurring in the sample collected from CETP.

In line and direction of the previously reported study and data on dyes and dye degradation, the present work was aimed to search for potential microbial populations occurring in the samples collected from the sites that have been contaminated by textile units wastes containing several dyes (as mixture) since last two to three decades.

In general, the objective of the present study was set forth to analyze the samples to explore microbial species having potential to decolorize and degrade dyes. The experimental work was designed to isolate, screen, and characterize (at least up to genus level) the organisms existing in the contaminated samples. It was also one of the main objectives to obtain information of the organisms on their efficiency to decolorize and degrade the dyes under laboratory conditions.

PHYSICO-CHEMICAL CHARACTERISTICS OF THE SAMPLES

Total five samples were collected from the common effluent treatment plant, located outside of Jetpur town, in autoclaved

BOD bottles for microbial analysis and in polythene bottles for physical and chemical analysis as describe in materials and methods before analysis. Analysis was carried out as per APHA (Andrew *et al.*, 1995).

When the samples were collected, the CETP was in working condition. The description of the five samples collected from the CETP is given in (Table 2.1). All the five samples were used for isolation of various bacterial populations. The five samples were analyzed for the selective parameters, such as, color, pH, COD and BOD.

Color comparison of the five samples did not show much variation, except that color changes from reddish green to light green in the treated samples (Stage-5). All the samples showed pH in the treated effluent, 7.6 to 8.8. However, the treated effluent showed slightly alkaline, *i.e.*, pH 7.6 (Table 3.1).

Sample	Stage of treatment	Color of effluents	pH	COD mg/L	BOD mg/L
Stag-1	Untreated-1	Reddish Green	8	580	295
Stag-2	Untreated-2	Greenish	8	560	290
Stag-3	Sludge free	Light green	9	533	315
Stag-4	Sludge	Light Green	8	520	350
Stag-5	Treated	Green	8	416	286

The value of COD estimated were ranging from 416 to 580 mg/L, *i.e.,* the lowest value, 416 mg/L was observed in the sample after treatment while, maximum value 580 mg/L for untreated samples.

Results of BOD analysis of the five samples indicate that, almost 50% COD was reduced during the treatment processes. It is also to be noted that domestic sewage of Jetpur town and the textile effluent are mixed before it is pumped to the treatment plant for further treatment.

Thus, the common domestic sewage microflora are mixed up before the wastewater drained into the treatment plant. The results, particularly, of COD and BOD had given a direction to design and study various microflora existing in the wastewater.

MICROBIOLOGICAL ANALYSIS OF THE SAMPLES

As stated earlier, the main goal of the present study was to systematically analyze the microbial populations that occur in the wastewater samples. To satisfy the aim, the following objectives were focused:

1. Biodiversity of microbial population of CETP
2. Enrichment and isolation of organisms and characterization
3. Screening of the bacterial isolates
4. Quantification of the bacterial isolates for their potentiality to decolorize the dyes
5. Selection of dyes and bacterial strains for detailed investigations
6. Optimization of cultural conditions that enhance the process of decolorization

Diversity of Microbial Population of CETP

The five samples collected at different treatment stages of CEPT were analyzed to find out the diversity of various microbial populations occurring in each stage.

By using the CMB medium, each sample inoculated and incubated, were transferred onto the nutrient agar plates and observed for the development of different colonies. The colonies developed were selected and transferred on nutrient agar slants to obtain pure cultures of the isolates and further examined for microscopic cellular morphology and Gram's reaction.

Thus, microbial observations of each sample for each stage were noted. The results obtained were presented graphically. (Fig. 3.2 & 3.3), as% distribution of Gram +ve, Gram –ve, actinomycetes, and yeast in each stage of the CETP. Total 50 microbial isolates were distinguished. (Fig. 3.2) reveals that Gram +ve organisms were dominating in the stage 1, 2, and

4 (59%, 52% and 64% respectively), whereas, Gram –ve dominates in the stages 3 and 5 (50% and 72% respectively). Gram –ve population was maximum (72%) in stage 5. Actinomycetes were absents in the stages 1, 2, 3, and 5, which was found 4% of the total microbial populations. Yeasts have been observed throughout the five stages. The results indicate the occurrence of several microbial floras in the CETP.

[Fig. 3.2] Numerical diversity of Gr+ve population at different treatment stages of textile effluents

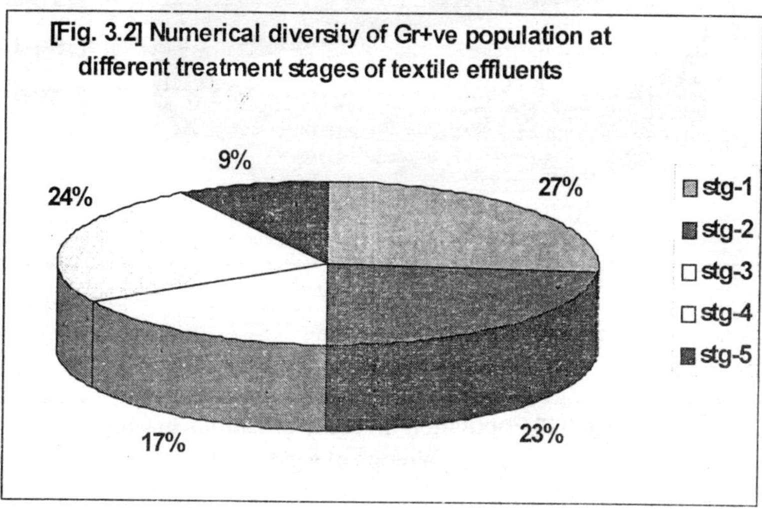

[Fig. 3.2] Numerical diversity of Gr-ve population at different treatment stages of textile effluents

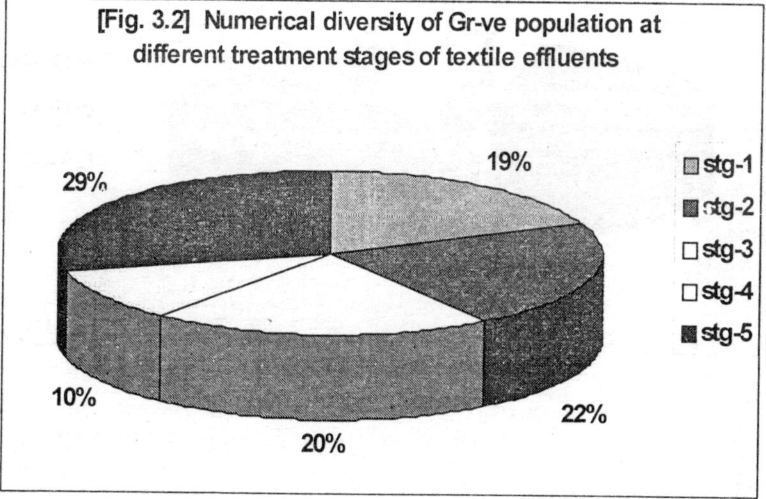

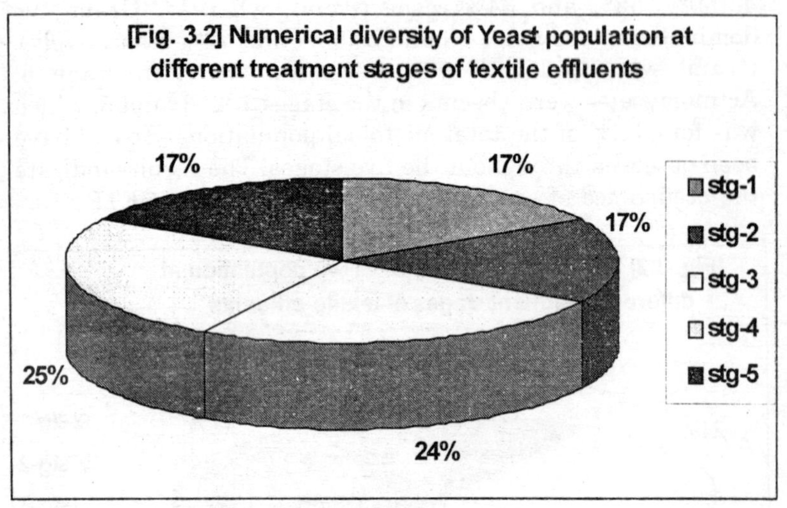
[Fig. 3.2] Numerical diversity of Yeast population at different treatment stages of textile effluents
17%
17%
17%
25%
24%
stg-1
stg-2
stg-3
stg-4
stg-5

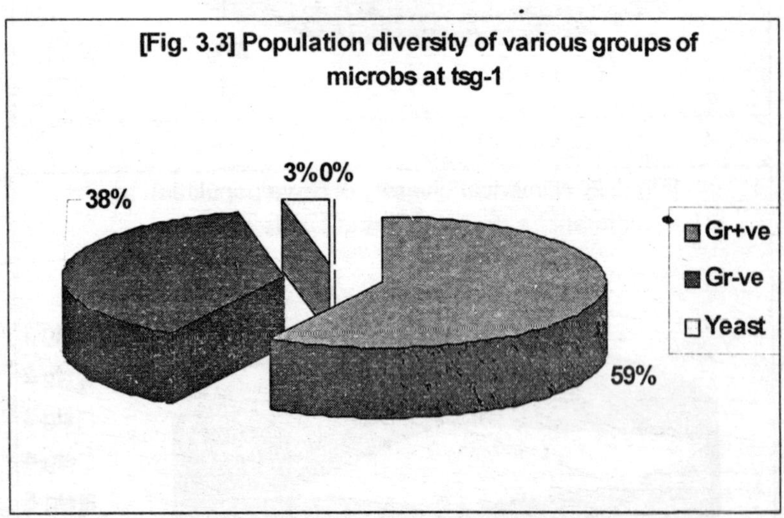
[Fig. 3.3] Population diversity of various groups of microbs at tsg-1
38%
3%0%
59%
Gr+ve
Gr-ve
Yeast

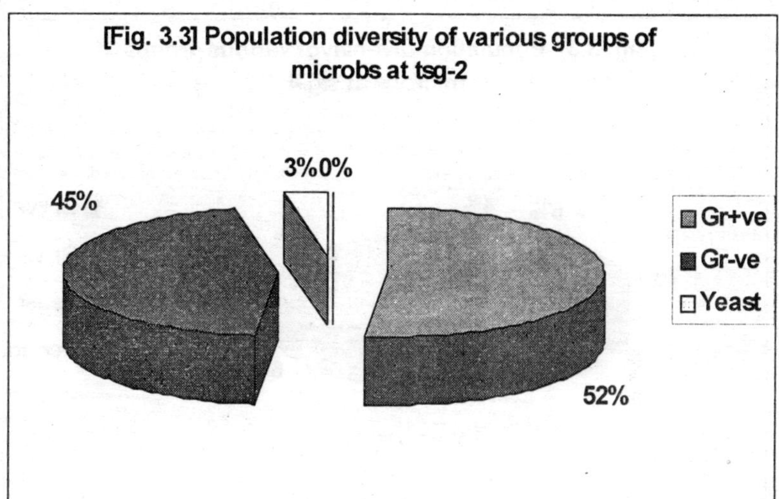
[Fig. 3.3] Population diversity of various groups of microbs at tsg-2
3%0%
45%
52%
Gr+ve
Gr-ve
Yeast

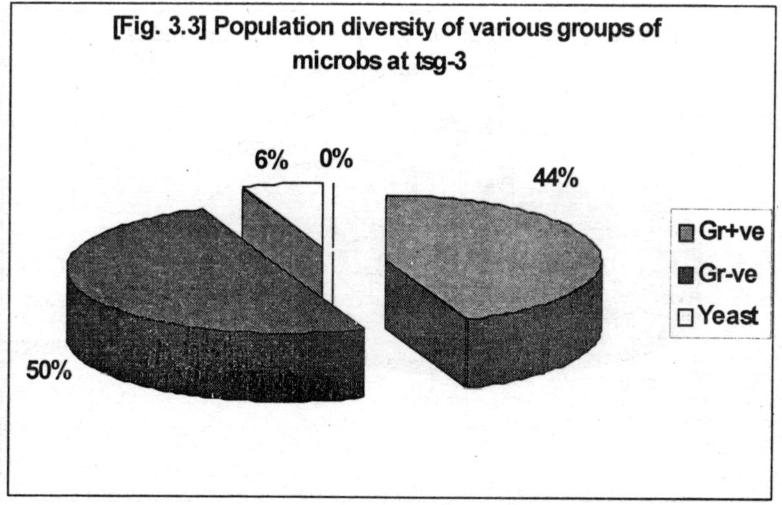
[Fig. 3.3] Population diversity of various groups of microbs at tsg-3
6%
0%
44%
50%
Gr+ve
Gr-ve
Yeast

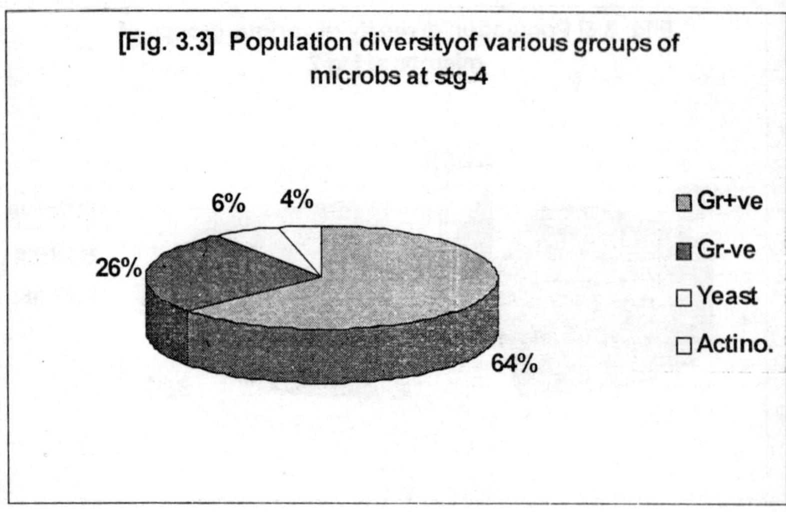
[Fig. 3.3] Population diversityof various groups of microbs at stg-4
6%
4%
26%
64%
Gr+ve
Gr-ve
Yeast
Actino.

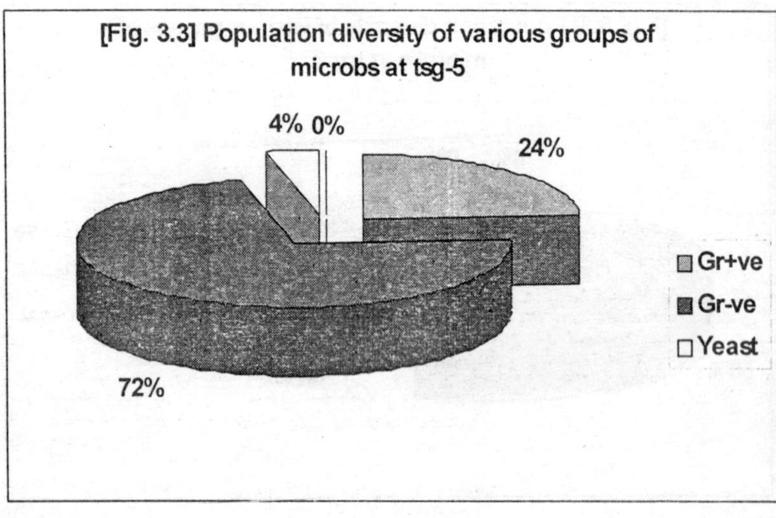
[Fig. 3.3] Population diversity of various groups of microbs at tsg-5
4% 0%
24%
72%
Gr+ve
Gr-ve
Yeast

Enrichment and Isolation of Dye Decolorizing Microbial Populations

To fulfill the objectives, a complex medium containing certain salts was employed. Since the beginning of the study for the isolation of bacterial populations, it was aimed to search for chemoheterotrophic organisms existing in the samples collected from the different stages of CETP. Therefore, the ingredients selected were peptone, Yeast extract, Magnesium sulfate, and phosphate salts, which would make the medium supplying C, N, S, P, and other growth factors needed by heterotrophs. Keeping in mind the search of a potential micobial population, this study was aimed to screen organisms having potentiality to decolorize various dyes, a solution of dyes (20) mixture (Table-1) was added in the medium to achieve selective isolation by enrichment of desired organisms.

CMB medium (100-150 ml) in conical flasks containing mixture of dyes were inoculated. Primary inoculum used was from the textile effluent samples, stored in sterile BOD bottle at 4°C. The flasks incubated at 30°C under two conditions, static and shaking, were examined at regular intervals for the disappearance of color in the flasks. From the selected flasks, loopful was transferred in the same fresh medium and incubated. The transfers were repeated three times from the culture flasks showed disappearance of color. Further selection of the bacterial populations was carried out by streaking a loopful from the culture flask on CMA (with dye mixture) plates.

The incubated plates (at 30°C) were examined after 24, 48, and 72 hours and colonies developed were selected and subcultured on the fresh medium. Thus, pure cultures obtained were transferred on agar slants having CMA medium, incubated and maintained at 4°C for further studies.

Enrichment media selectively encourage the growth of specific bacteria occurring in mixed populations. The solutions of dye mixture added in the medium considered as an selective agent and criterion chosen to examine the growth of desired organisms was the disappearance of color of the medium. Thus, a visible distinction of the medium color in the flask facilitated selection.

From the initiation of the isolation experiments, it was distinctly noticed that those flasks incubated under shaking condition did not show visibly distinguishable color compared to the control, and therefore, further studies were conducted only under stationary batch culture.

Identification of the Bacterial Isolates

These thirty bacterial isolates were considered for their characterization, based on colony characteristics, Gram's reaction and cell morphology, and growth patterns in nutrient broth, and biochemical tests (Table 3.2, 3.3, 3.4, 3.5, 3.5.1, 3.5.2, 3.5.3, 3.5.4, 3.5.5, and 3.6).

Identification based on various characters studied are given in Table 3.7. From the analyzed characters, the isolates were tentatively grouped into various generic level (Bergey's Manual, 1994). The bacterial isolates were arranged (Table-3.7) show that many of them occur very commonly in such environment. They all belonged to heterotrophic group.

Screening of Bacterial Isolates

The bacterial isolates obtained during the enrichment process were numbered as kot-01 to kot-55. These 55 isolates were further screened on the basis of decolorization of the dye mixture solution previously used. The procedure described in materials and methods, an activated inoculum of each isolate was prepared in CMB medium and inoculated in 250 conical flasks containing 150 ml of CMB + dye mixture. The flasks incubated were examined at the interval of 24 hours.

After 24 hours of incubation at 30°C, flasks were examined for the decolorization of dyes in the medium and compared with the control (CMB + Dye mixture, but without inoculum). The observations were continued for 72 hours, at 24 hours interval. The selection criteria of the bacterial isolates were based on the time taken (24, 48, and 72 hours) to decolorize dye mixture. Thus, from the primary screening only thirty isolates that were recoreded to be capable to decolorize mixture of dyes completely within 48 hours were selected.

TABLE: 3.2

RESULTS OF GROWTH ON NUTRIENT AGAR PLATES SHOWING COLONY CHARACTERISTICS AND GRAM'S REACTION

Culture No.	COLONY CHARACTERISTICS							Gram's reaction
	Size	Pigmentation	Form	Margin	Opacity	Elevation	Texture	
Kot-1	Moderate	Gray	Circular (Unbroken peripheral edge)	Entire (sharply defined)	Opaque	Convex (Dome shaped elevation)	Smooth	Gr-ve small rods with round ends, occurring singly
Kot-2	Large	Gray	Circular (Unbroken peripheral edge)	Entire (Even)	Opaque	Convey (Dome shaped elevation)	Smooth	Gr-ve small rods with round ends, occurring singly
Kot-3	Small	Yellowish	Circular (Unbroken peripheral edge)	Entire (sharply defined)	Opaque	Convey	Smooth	Gr-ve small, rod, occurring singly
Kot-4	Moderate	Gray	Circular (Unbroken peripheral edge)	Entire (sharply defined)	Opaque	Umbonate (Raised with elevated central region)	Smooth	Gr+ve big rods with round ends & terminal spore, occurring in chain & singly
Kot-5	Large	White	Rhizoid (Root like spreading grown)	Filamentous (thread like spreading edge)	Opaque	Flat (elevation not discernible)	Smooth	Gr+ve big rods with round ends & occurring in chain

TABLE: 3.2

RESULTS OF GROWTH ON NUTRIENT AGAR PLATES SHOWING COLONY CHARACTERISTICS AND GRAM'S REACTION

Culture No.	COLONY CHARACTERISTICS							Gram's reaction
	Size	Pigmentation	Form	Margin	Opacity	Elevation	Texture	
Kot-6	Large	White	Irregular (Indented peripheral edge)	Lobate (Marked Indentations)	Opaque	Flat (Elevation net discernible)	Rough	Gr-ve small rods with pointed end, but less sharp occurring singly & in clusters
Kot-7	Large	White	Circular (Unbroken peripheral edge)	Lob ate (marked indentation)	Opaque	Umbonate (Raised)	Rough	Gr-ve cocci (Tetrads)
Kot-8	Large	Gray	Circular (Unbroken peripheral edge)	Serrate (Tooth like appearance)	Opaque	Flat (Elevation net discernible)	Smooth	Gr-ve big rods, with round ends, occurring in chain & in pair
Kot-9	Moderate	(Gray)	Circular	Entire (Even)	Opaque	Umbonate (Raised)	Rough	Gr+ve big rods with round ends, occurring singly & in cluster
Kot-10	Moderate	Light yellow	Circular (Unbroken peripheral edge)	Entire (sharply defined)	Transparent	Convex (Domed shaped Elevation)	Smooth	Gr-ve small rods with round ends occurring single

TABLE: 3.2

RESULTS OF GROWTH ON NUTRIENT AGAR PLATES SHOWING COLONY CHARACTERISTICS AND GRAM'S REACTION

Culture No.	COLONY CHARACTERISTICS							Gram's reaction
	Size	Pigmentation	Form	Margin	Opacity	Elevation	Texture	
Kot-11	Moderate	White	Circular (Unbroken peripheral edge)	Entire (Even)	Opaque	Flat (Elevation discernible)	Smooth	Gr-ve small rods with blunt ends occurring single
Kot-12	Small	Light yellow	Circular (Unbroken peripheral edge)	Entire (Even)	Transparent	Raised (slightly elevated)	Smooth	Gr-ve small rods with round ends occurring single
Kot-13	Moderate	Gray	Circular	Entire (sharply defined)	Opaque	Convex (domed shaped elevation)	Smooth	Gr+ve small rods with round ends occurring single
Kot-14	Moderate	White	Rhizoid (Root like spreading growth)	Filamentous (Thread like spreading edge)	Opaque	Raised (Slightly elevated)	Smooth	Gr+ve big rods with round ends, with terminal spare, singly
Kot-15	Moderate	White	Rhizoid (Root like spreading growth)	Filamentous (Thread like spreading growth)	Opaque	Raised (Slightly elevated)	Smooth	Gr-ve small rods with round ends occurring singly & double

TABLE: 3.2

RESULTS OF GROWTH ON NUTRIENT AGAR PLATES SHOWING COLONY CHARACTERISTICS AND GRAM'S REACTION

Culture No.	COLONY CHARACTERISTICS							Gram's reaction
	Size	Pigmentation	Form	Margin	Opacity	Elevation	Texture	
Kot-16	Moderate	Gray	Rhizoid	Filamentous	Opaque	Flat	Smooth	Gr+ve small rods with round ends occurring singly
Kot-17	Small	White	Circular	Entire	Opaque	Flat	Smooth	Gr+ve small rods with round ends occurring singly
Kot-18	Moderate	White	Irregular (Indented peripheral edge)	Undulate (Wavy indentation)	Opaque	Raised (Slightly raised)	Smooth	Gr+ve big rods with blunt ends, occurring singly
Kot-19	Small	White	Rhizoid (Root like spreading growth)	Filamentous (Thread like spreading edge)	Opaque	Flat (Elevation net discernible)	Smooth	Gr+ve small rods with round ends occurring singly
Kot-20	Moderate	Gray	Rhizoid	Lobate	Opaque	Raised	Smooth	Gr+ve small rods with round ends occurring singly & in chains

TABLE: 3.2

RESULTS OF GROWTH ON NUTRIENT AGAR PLATES SHOWING COLONY CHARACTERISTICS AND GRAM'S REACTION

Culture No.	COLONY CHARACTERISTICS							Gram's reaction
	Size	Pigmentation		Margin	Opacity	Elevation	Texture	
Kot-21	Moderate	Gray	Circular	Entire	Opaque	Convex	Smooth	Gr+ve small rods with round ends occurring singly.
Kot-22	Big	White	Circular	Serrate	Opaque	Convex	Rough	Gr+ve big rods with round ends, occurring in chain with terminal spore
Kot-23	Small	White	Circular	Entire	Slight transparent	Raised	Smooth	Gr-ve small rods with round ends occurring singly
Kot-23	Moderate	White	Circular	Entire	Opaque	Raised	Smooth	Gr-ve small rods with round ends occurring singly
Kot-24	Big	Gray	Irregular	Serrate	Opaque	Raised	Smooth	Gr+ve small rods with round ends occurring singly

TABLE: 3.2

RESULTS OF GROWTH ON NUTRIENT AGAR PLATES SHOWING COLONY CHARACTERISTICS AND GRAM'S REACTION

Culture No.	COLONY CHARACTERISTICS							Gram's reaction
	Size	Pigmentation	Form	Margin	Opacity	Elevation	Texture	
Kot-25	Spreading growth	Gray	Rhizoid	Filamentous	Opaque	Flat	Smooth	Gr+ve big rods, occurring in clusters
Kot-26	Small	Gray	Rhizoid	Filamentous	Opaque	Umbonate	Rough	Gr-ve small rods with round ends occurring singly
Kot-27	Small	Dirty white	Rhizoid	Filamentous	Opaque	Raised	Rough	Gr+ve small rods with round ends occurring singly
Pseu-I	Small	Yellowish	Circular (Unbroken peripheral edge)	Entire (sharply defined)	Transparent	Raised (Slightly elevated)	Smooth	Gr-ve small rods with pointed end occurring singly
Pseu-II	Moderate	Yellowish	Circular	Entire (sharply defined)	Dull Transparent	Raised (Slightly elevated)	Smooth	Gr-ve small rods with pointed end but less sharp occurring singly
Pseu-III	Big	Yellowish	Circular (Broken peripheral edge)	Entire (sharply defined)	Transparent (Glowing like silver)	Raised (Slightly elevated)	Smooth	Gr-ve small rods with pointed end but less sharp occurring singly

TABLE 3.3 : RESULTS OF GROWTH BEHAVIOUR ON NUTRIENT AGAR SLANTS

Culture No.	Abundance of Growth	Growth Characteristics Pigmentation	Optical Charac-teristics	Form
Kot-1	Moderate	Gray	Opaque	Filiform
Kot-2	Large	Gray	Opaque	Filiform
Kot-3	Moderate	Yellowish	Transparent	Filiform
Kot-4	Moderate	Gray	Opaque	Filiform
Kot-5	Large	White	Opaque	Arborescent
Kot-6	Large	White	Opaque	Echinulate
Kot-7	Large	White	Opaque	Echinulate
Kot-8	Large	Gray	Opaque	Echinulate
Kot-9	Moderate	Gray	Opaque	Filiform
Kot-10	Moderate	Yellowish	Transparent	Filiform
Kot-11	Large	White	Opaque	Filiform
Kot-12	Moderate	Yellowish	Transparent	Filiform
Kot-13	Large	Gray	Opaque	Filiform
Kot-14	Moderate	White	Opaque	Rhizoid
Kot-15	Large	White	Opaque	Rhizoid
Kot-16	Large	Gray	Opaque	Rhizoid
Kot-17	Moderate	White	Opaque	Bended
Kot-18	Large	Gray	Opaque	Arborescent
Kot-19	Moderate	White	Opaque	Rhizoid
Kot-20	Large	Gray	Opaque	Rhizoid
Kot-21	Large	Gray	Opaque	Filiform
Kot-22	Large	White	Opaque	Effuse
Kot-23	Large	White	Opaque	Filiform
Kot-24	Moderate	White	Opaque	Filiform
Kot-25	Large	Gray	Opaque	Rhizoid
Kot-26	Large	Gray	Opaque	Arborescent
Kot-27	Moderate	Gray	Opaque	Rhizoid
Pseu-1	Small	Yellowish	Transparent	Filiform
Pseu-ll	Moderate	Yellowish	Transparent	Filiform
Pseu-lll	Big	Yellowish	Transparent	Filiform

TABLE: 3.4

RESULTS OF GROWTH PATTERN IN NUTRIENT BROTH

Culture No.	GROWTH CHARACTERISTICS			
	Turbidity	Flocculants	Pellicle	Deposit
Kot-1	Finely dispersed growth through	-	-	-
Kot-2	Finely dispersed growth through	-	-	-
Kot-3	Finely dispersed growth through	-	-	-
Kot-4	Finely dispersed growth through	-	-	-
Kot-5	Finely dispersed growth through	-	-	-
kot-6	Finely dispersed growth through	-	-	-
Kot-7	-	Flaky aggregate dispersed throughout	-	-
Kot-8	-	Flaky aggregate dispersed throughout	-	-
Kot-9	-	-	-	Concentration of growth at the bottom of broth culture
Kot-10	-	-	Thick, Pod like growth on surface	-
Kot-11	Finely dispersed growth throughout	-	-	-
Kot-12	Finely dispersed growth throughout	Flaky aggregate dispersed throughout	-	-
Kot-13	-	-	-	Concentration of growth at the bottom of broth culture

TABLE: 3.4

RESULTS OF GROWTH PATTERN IN NUTRIENT BROTH

Culture No.	GROWTH CHARACTERISTICS			
	Uniform fine Turbidity	Flocculants	Pellicle	Sediment
Kot-14	-	-	-	Concentration of growth at the bottom of broth culture
Kot-15	Finely dispersed growth throughout	-	-	-
Kot-16	Finely dispersed growth throughout	-	-	-
Kot-17	Finely dispersed growth throughout	-	-	-
Kot-18	Finely dispersed growth throughout	-	-	-
Kot-19	Finely dispersed growth throughout	-	-	-
Kot-20	Finely dispersed growth throughout	-	-	-
Kot-21	Finely dispersed growth throughout	-	-	-
Kot-22	-	Flaky aggregate dispersed throughout	-	-
Kot-23	-	Flaky aggregate dispersed throughout	-	-
Kot-24	Finely dispersed growth throughout	-	-	-

TABLE: 3.4

RESULTS OF GROWTH PATTERN IN NUTRIENT BROTH

Culture No.	GROWTH CHARACTERISTICS			
	Uniform fine Turbidity	Flocculants	Pellicle	Sediment
Kot-25	-	-	Thick padlike growth on surface	-
Kot-26	-	-	-	Concentration of growth at the bottom of broth culture
Kot-27	-	-	-	Concentration of growth at the bottom of broth culture
Pseu-I	Finely dispersed growth through	-	-	-
Pseu-II	Finely dispersed growth through	-	-	-
Pseu-III	Finely dispersed growth through	-	-	-

TABLE: 3.5.1

RESULTS SHOWING BIOCHEMICAL TESTS-FERMENTATION OF VARIOUS SUGARS

Bacterial isolates	Fermentation							
	Glc	Mal	Suc	Lac	Man	Xyl	Ino	Sorb
Kot-01	+/+	+/+	+/+	-/-	+/+	+/+	-/-	+/-
Kot-02	+/-	+/-	+/-	+/-	-/-	-/-	-/-	-/-
Kot-03	+/-	+/-	+/-	+/-	+/-	+/-	+/-	+/-
Kot-04	+/+	+/+	-/-	-/-	+/+	+/+	-/-	+/+
Kot-05	+/-	+/-	+/-	+/-	-/-	-/-	-/-	-/-
Kot-06	+/-	+/-	+/-	-/-	+/-	+/-	-/-	-/-
Kot-07	+/-	+/-	+/-	-/-	+/-	+/-	-/-	-/-
Kot-08	+/-	+/-	+/-	-/-	-/-	-/-	-/-	-/-
Kot-09	+/-	+/-	-/-	-/-	-/-	-/-	-/-	-/-
Kot-10	+/-	+/-	+/-	+/-	-/-	-/-	-/-	-/-
Kot-11	+/-	+/-	+/-	-/-	+/-	-/-	-/-	-/-
Kot-12	+/-	+/-	+/-	-/-	+/-	+/-	-/-	-/-
Kot-13	+/+	+/+	+/+	-/-	+/+	+/+	-/-	-/-
Kot-14	+/-	+/-	+/-	+/-	-/-	-/-	-/-	-/-
Kot-15	+/-	+/-	+/-	-/-	-/-	-/-	-/-	-/-
Kot-16	+/-	+/-	+/-	+/-	-/-	-/-	-/-	-/-
Kot-17	+/-	-/-	+/-	-/-	-/-	-/-	-/-	-/-
Kot-18	+/-	+/-	+/-	-/-	-/-	-/-	-/-	-/-
Kot-19	+/-	-/-	+/-	-/-	+/-	-/-	-/-	-/-
Kot-20	+/-	+/-	+/-	+/-	-/-	-/-	-/-	-/-
Kot-21	+/+	+/+	+/+	+/+	+/+	+/+	-/-	+/+
Kot-22	+/+	+/-	+/-	+/-	-/-	-/-	-/-	-/-
Kot-23	+/-	+/+	+/+	+/+	+/+	+/+	+/+	+/+
Kot-24	+/-	+/-	+/-	-/-	+/-	+/-	-/-	-/-
Kot-25	+/-	+/-	+/-	+/-	-/-	-/-	-/-	-/-
Kot-26	+/-	+/-	+/-	-/-	+/-	+/-	-/-	-/-
Kot-27	+/-	+/-	+/-	+/-	-/-	-/-	-/-	-/-
Pseu-I	+/-	+/-	+/-	-/-	-/-	-/-	-/-	-/-
Pseu-II	-/-	-/-	-/-	-/-	-/-	-/-	-/-	-/-
Pseu-III	-/-	-/-	-/-	-/-	-/-	-/-	-/-	-/-

[**+/+** means acid production/ gas production; **-/-** means no acid/no gas production]

[**Glc**: Glucose; **Mal**: Maltose; **Sur**: Sucrose; **Lac**: Lactose; **Man**: Manitol; **Xyl**: Xylose; **Ino**: Inositol; **Sorb**: Sorbotol]

TABLE: 3.5.2

RESULTS SHOWING BIOCHEMICAL TEST-HYDROLYSIS OF DIFFERENT COMPLEX POLYMERIC BIOMOLECULES

Bacterial isolates	Hydrolysis of				Utilization of
	Casein	Gelatin	Starch	Tributyrin	Citrate
Kot-01	-	+	-	+	+
Kot-02	-	+	-	+	+
Kot-03	+	+	-	+	-
Kot-04	+	+	-	+	+
Kot-05	+	+	-	+	-
Kot-06	+	+	+	+	+
Kot-07	+	+	+	+	+
Kot-08	-	+	-	+	+
Kot-09	+	+	+	+	-
Kot-10	-	+	-	+	+
Kot-11	-	+	+	+	+
Kot-12	+	+	-	+	+
Kot-13	-	-	+	+	+
Kot-14	-	+	-	+	+
Kot-15	+	+	+	+	-
Kot-16	+	+	+	+	-
Kot-17	+	+	+	+	-
Kot-18	+	+	-	-	+
Kot-19	-	-	+	+	-
Kot-20	-	-	+	-	+
Kot-21	-	-	-	-	-
Kot-22	+	+	-	+	+
Kot-23	-	-	-	-	+
Kot-24	-	+	-	+	+
Kot-25	+	+	-	-	-
Kot-26	+	+	+	+	+
Kot-27	+	-	+	-	+
Pseu-I	+	+	-	+	+
Pseu-II	+	+	-	+	+
Pseu-III	+	+	-	+	+

[Where, (+) means hydrolysis or utilization, (-) means not hydrolysis or utilization]

TABLE: 3.5.3
RESULTS SHOWING BIOCHEMICAL TESTS (CONFIRMATIVE)

Bacterial isolates	Tests				
	Methyl red	Voges-Prokaues	Catalase	Oxidase	Motility
Kot-01	-	-	+	-	+
Kot-02	-	-	+	+	+
Kot-03	-	-	+	+	+
Kot-04	-	-	+	+	+
Kot-05	-	-	-	+	+
Kot-06	-	-	+	+	+
Kot-07	-	-	+	+	+
Kot-08	-	-	+	-	+
Kot-09	-	-	+	+	+
Kot-10	-	+	+	+	+
Kot-11	-	-	+	+	+
Kot-12	-	-	+	+	+
Kot-13	-	-	+	+	+
Kot-14	-	-	+	+	+
Kot-15	-	-	+	-	+
Kot-16	-	-	+	+	+
Kot-17	-	-	+	+	+
Kot-18	-	-	+	+	+
Kot-19	-	-	+	+	+
Kot-20	-	-	+	+	+
Kot-21	-	+	+	+	+
Kot-22	-	+	+	+	+
Kot-23	-	+	+	+	+
Kot-24	-	-	+	+	+
Kot-25	-	-	+	+	+
Kot-26	-	-	+	-	+
Kot-27	-	-	+	+	+
Pseu-I	-	-	+	+	+
Pseu-II	-	-	+	+	+
Pseu-III	-	-	+	+	+

[Where, (+) means hydrolysis or utilization, (-) means not hydrolysis or utilization]

TABLE: 3.5.4

RESULTS SHOWING BIOCHEMICAL TESTS (TSI)

Bacterial Isolates	Triple Sugar Iron Slant			
	Slant	Butt	Gas	H_2S
Kot-01	R	Y	+	-
Kot-02	Y	Y	-	-
Kot-03	Y	Y	-	-
Kot-04	R	Y	-	-
Kot-05	Y	Y	-	-
Kot-06	Y	R	-	-
Kot-07	R	Y	-	-
Kot-08	R	Y	-	-
Kot-09	Y	Y	-	-
Kot-10	R	Y	-	-
Kot-11	R	R	+	-
Kot-12	R	Y	-	-
Kot-13	R	Y	+	-
Kot-14	R	Y	-	-
Kot-15	R	Y	-	-
Kot-16	R	Y	-	-
Kot-17	Y	Y	-	-
Kot-18	R	Y	-	-
Kot-19	Y	R	-	-
Kot-20	R	Y	-	-
Kot-21	R	Y	+	-
Kot-22	R	Y	-	-
Kot-23	Y	Y	+	-
Kot-24	R	Y	-	-
Kot-25	R	Y	-	-
Kot-26	R	Y	-	-
Kot-27	Y	Y	-	-
Pseu-I	R	Y	-	-
Pseu-II	R	R	-	-
Pseu-III	R	Y	-	-

[R, means, Red slant, Y means, Yellow butt, (+) means, Gas produced (-) means, gas not produced]

TABLE: 3.5.5

RESULTS SHOWING UTILIZATION OF VARIOUS NITROGENOUS COMPOUNDS

Bacterial isolates	TESTS				
	Indole production	H_2S production	Ammonia production	NO_3 reduction	Deamination test
Kot-01	-	-	+	-	-
Kot-02	-	-	+	-	-
Kot-03	-	-	+	+	-
Kot-04	+	-	+	-	-
Kot-05	-	-	+	-	-
Kot-06	-	-	+	-	-
Kot-07	-	-	+	-	-
Kot-08	-	-	+	-	-
Kot-09	-	-	+	-	-
Kot-10	-	+	+	-	-
Kot-11	-	-	+	-	-
Kot-12	-	-	+	-	-
Kot-13	+	-	+	-	-
Kot-14	-	-	+	-	-
Kot-15	-	-	+	+	-
Kot-16	-	-	+	-	-
Kot-17	-	+	+	-	-
Kot-18	-	-	+	-	-
Kot-19	-	-	+	-	-
Kot-20	-	-	+	-	-
Kot-21	+	-	+	-	-
Kot-22	-	-	+	-	-
Kot-23	-	-	+	-	-
Kot-24	-	-	+	-	-
Kot-25	-	-	+	-	-
Kot-26	-	-	+	-	-
Kot-27	-	-	+	-	-
Pseu-I	-	-	+	-	+
Pseu-II	-	-	+	-	+
Pseu-III	-	-	+	-	+

+) means, test is positive, (-) means, test is negative]

TABLE: 3.6
RESULTS SHOWING GROWTH AT DIFFERENT TEMPERATURE

Bacterial isolates	Growth temperature(^{0}C)				
	25	37	40	45	50
Kot-01	+	+	+	+	-
Kot-02	+	+	+	+	-
Kot-03	+	+	+	+	-
Kot-04	+	+	+	+	+
Kot-05	+	+	+	+	+
Kot-06	+	+	+	+	-
Kot-07	+	+	+	+	-
Kot-08	+	+	+	+	-
Kot-09	+	+	+	+	+
Kot-10	+	+	+	+	-
Kot-11	+	+	+	+	-
Kot-12	+	+	+	+	-
Kot-13	+	+	+	+	+
Kot-14	+	+	+	+	+
Kot-15	+	+	+	+	-
Kot-16	+	+	+	+	+
Kot-17	+	+	+	+	+
Kot-18	+	+	+	+	+
Kot-19	+	+	+	+	+
Kot-20	+	+	+	+	+
Kot-21	+	+	+	+	+
Kot-22	+	+	+	+	+
Kot-23	+	+	+	+	-
Kot-24	+	+	+	+	+
Kot-25	+	+	+	+	+
Kot-26	+	+	+	+	-
Kot-27	+	+	+	+	+
Pseu-I	+	+	+	+	+
Pseu-II	+	+	+	+	+
Pseu-III	+	+	+	+	+

[Where, (+) means, growth was observed,
and (-) means, growth was not observed]

It is evident from the results recorded that, all 30 bacterial strains were more or less capable to reduce color in the medium containing dyes when cultivated under stationary condition. There was no visible reduction of color of the medium grown under shaking condition.

Table 3.8 shows results of cultures grown under static conditions, *i.e.*, lowest value was 39% (kot-16) and highest value 72% (Pseu-I), indicating that all the 30 cultures have more or less potential to reduce color.

TABLE: 3.7 LIST OF TENTATIVELY IDENTIFIED BACTERIAL ISOLATES

Culture No.	Name of isolates	Culture No.	Name of isolates
Kot-01	*Acetobacter sp.*	Kot-16	*Bacillus sp.*
Kot-02	*Pseudomonas sp.*	Kot-17	*Bacillus sp.*
Kot-03	*Azotobacter sp.*	Kot-18	*Bacillus sp.*
Kot-04	*Bacillus sp.*	Kot-19	*Bacillus sp.*
Kot-05	*Bacillus sp.*	Kot-20	*Bacillus sp.*
Kot-06	*Legionella sp.*	Kot-21	*Bacillus sp.*
Kot-07	*Paracoccus sp.*	Kot-22	*Bacillus sp.*
Kot-08	*Enterobacter sp.*	Kot-23	*Gluconobacter sp.*
Kot-09	*Bacillus sp.*	Kot-24	*Bacillus sp.*
Kot-10	*Xanthomonas sp.*	Kot-25	*Bacillus sp.*
Kot-11	*Agrobacter sp.*	Kot-26	*Phyllobacterium sp.*
Kot-12	*Xanthomonas sp.*	Kot-27	*Bacillus sp.*
Kot-13	*Bacillis sp.*	Pseu-I	*Pseudomonas sp.*
Kot-14	*Bacillus sp.*	Pseu-II	*Pseudomonas sp.*
Kot-15	*Azotobacter sp.*	Pseu- III	*Pseudomonas sp.*

TABLE: 3.8

DECOLORIZATION OF DYES BY SELECTED BACTERIAL ISOLATES UNDER STATIC CONDITION AFTER 72 HOURS

Isolate	Dyes decolorized (%)									
	Jd 00	Jd 01	Jd 02	Jd 03	Jd 04	Jd 05	Jd 06	Jd 07	Jd 08	Jd 10
Kot-01	86	72	85	67	56	72	38	66	66	66
Kot-02	70	61	85	63	54	69	39	59	56	65
Kot-03	87	74	66	64	47	68	51	64	57	62
Kot-04	73	67	76	70	55	72	49	62	68	61
Kot-05	98	92	85	64	61	64	48	53	61	68
Kot-06	84	67	81	69	52	65	52	68	64	59
Kot-07	89	72	81	62	55	62	47	66	65	75
Kot-08	93	78	80	65	53	68	48	67	59	62
Kot-09	94	90	80	63	52	63	41	65	58	53
Kot-10	92	84	70	64	50	62	59	59	63	26
Kot-11	68	59	58	61	53	52	49	62	56	66
Kot-12	54	65	55	59	60	64	58	59	65	55
Kot-13	65	58	57	64	51	53	65	49	56	55
Kot-14	51	53	59	61	63	58	54	52	45	46
Kot-15	63	59	51	61	53	52	49	48	51	42
Kot-16	41	39	38	46	37	45	42	33	39	37
Kot-17	55	45	56	47	42	39	52	38	53	41
Kot-18	43	39	30	33	31	35	40	39	34	38
Kot-19	60	61	59	51	49	63	66	67	54	59
Kot-20	50	55	45	54	44	49	48	51	58	61
Kot-21	49	39	45	54	38	39	50	51	49	58
Kot-22	52	53	45	55	49	47	59	40	51	57
Kot-23	54	49	48	58	46	51	40	58	56	51
Kot-24	65	61	59	69	70	63	58	64	67	59
Kot-25	56	38	39	41	54	48	49	39	51	41
Kot-26	59	49	58	48	57	55	59	53	54	44
Kot-27	60	53	61	54	53	57	55	54	61	63
Pseu-I	95	91	85	65	61	58	48	49	39	55
Pseu-II	97	96	90	61	66	51	49	52	35	51
Pseu-III	94	92	40	54	62	56	51	53	71	24

Poor decolorization : Below 50 %
Medium decolorization : Between 50 % to 75 %
Highly potential : Above 75 %

TABLE: 3.8

DECOLORIZATION OF DYES BY SELECTED BACTERIAL ISOLATES UNDER STATIC CONDITION

Isolate	Dyes decolorized (%)									
	Jd 12	**Ad 13**	**Ad 14**	**Ad 15**	**Ad 16**	**Ad 17**	**Ad 19**	**Kd 22**	**Kd 23**	**Kd 24**
Kot-01	49	66	32	47	81	51	65	49	55	50
Kot-02	47	65	41	38	71	46	69	38	59	54
Kot-03	71	51	25	46	82	49	58	47	61	48
Kot-04	57	56	35	39	77	58	62	38	68	70
Kot-05	56	64	39	52	58	57	53	32	63	69
Kot-06	65	52	42	56	82	66	64	31	58	62
Kot-07	45	54	46	55	75	63	67	44	59	68
Kot-08	57	65	54	68	80	58	58	45	49	48
Kot-09	59	35	55	71	61	54	62	56	58	59
Kot-10	65	68	58	70	82	59	59	52	57	37
Kot-11	61	61	59	49	61	63	65	56	51	58
Kot-12	56	48	49	59	53	52	55	49	53	60
Kot-13	61	59	49	53	62	58	49	48	56	51
Kot-14	49	53	49	41	39	45	47	42	51	44
Kot-15	61	52	54	53	61	59	57	48	52	46
Kot-16	37	39	38	41	36	41	43	39	37	33
Kot-17	55	53	49	61	48	57	46	55	44	41
Kot-18	44	51	49	53	52	42	48	46	44	42
Kot-19	60	61	65	64	68	67	69	62	63	60
Kot-20	55	51	54	59	49	48	43	46	45	42
Kot-21	49	47	57	53	51	57	59	58	60	48
Kot-22	51	49	53	51	47	40	46	45	43	41
Kot-23	58	60	68	64	63	61	59	54	56	51
Kot-24	67	64	61	63	68	69	67	59	57	61
Kot-25	57	56	60	63	62	61	62	59	52	62
Kot-26	58	59	51	53	57	55	56	61	60	58
Kot-27	60	69	65	63	65	59	57	55	51	53
Pseu-I	88	78	49	86	82	88	85	86	86	87
Pseu-II	83	81	37	83	86	80	69	78	84	82
Pseu-III	79	88	50	87	71	87	77	89	83	87

Poor decolorization	: Below 50 %
Medium decolorization	: Between 50 % to 75 %
Highly potential	: Above 75 %

It is also quite evident from (Table 3.8) that there were significant variations to decolorize the 20 dyes by individual bacterial isolates. Some of the dyes (Jd 00, Jd 01, and Jd 04) were effectively decolorized by all the 30 isolates, within 72 hours of incubation. The experiment conducted was to find out the potential isolates that decolorize the 20 dyes. From the decolorizing assay, it was decided to use only the selected 13 bacterial strains (isolates *i.e.,* kot 01 to kot 10, and Pseu-I, Pseu-II, and Pseu-III) for further screening.

ANTIBIOTIC SENSITIVITY ASSAY

Table 3.9.1, 3.9.2, 3.9.3, and 3.9.4 show the sensitivities of various bacterial isolates (Kot-01 to Kot-30) against 28 antibiotics. The diameter of each inhibition zone was considered as a measure of antibiotic sensitivity of the isolate. The results obtained for the sensitivity tests were compared with the standard data given in the table of the HiMedia Manual (1998). The antibiotics sensitivity determination made to generate and maintain records of the isolates for future studies.

SECONDARY SCREENING OF THE BACTERIAL STRAINS

The previous experiments showed that the 13 strains were decolorizing effectively only three dyes, *i.e.,* Kemifix Red F6B. RS Red H5BL, and RS Brown HGRL.

Young cultures (2 ml, 6 hours young having approx. 1.0 OD) of each of the thirteen bacterial strains were inoculated in the CMB medium containing 0.02% of individual dyes and incubated at 30°C in both stationary and shake flask conditions. Fig. 3.4, 3.5, and 3.6 show results of the performance of each bacterial strain with the individual dye, in terms of% decolorization at 24, 48, and 72 hours, in static and shake culture flasks. Kot-5, Pseu-I, and Pseu-II bacterial strains were most efficient as 90% or above 90% decolorization of dye Kemifix Red F6B observed within 24 hours, under static condition, whereas the shake condition did not show any significant decolorization even at 72 hour interval.

TABLE: 3.9.1

ANTI-MICROBIAL ACTIVITY OF VARIOUS BACTERIAL ISOLATES

Sr.No.	Culture No.	Antibiotics Tested							
		Ch	Cd	Co	E	G	Of	P	Va
01	Kot-01	08	24	00	08	16	20	00	12
02	Kot-02	08	22	00	10	14	18	08	12
03	Kot-03	32	16	28	18	20	24	28	18
04	Kot-04	22	18	24	16	18	22	26	16
05	Kot-05	10	20	00	12	18	22	10	14
06	Kot-06	08	20	00	12	18	18	08	12
07	Kot-07	24	22	20	08	20	24	22	18
08	Kot-08	10	30	00	16	18	24	12	14
09	Kot-09	08	20	00	18	09	22	20	00
10	Kot-10	12	24	20	26	10	20	00	00
11	Kot-11	20	20	08	20	00	18	22	00
12	Kot-12	20	26	22	00	22	30	18	00
13	Kot-13	30	28	10	00	02	10	18	00
14	Kot-14	24	26	12	24	26	28	30	18
15	Kot-15	10	22	25	08	22	25	00	20
16	Kot-16	15	20	24	26	25	24	03	00
17	Kot-17	17	24	26	09	23	27	00	08
18	Kot-18	20	28	10	10	10	22	20	09
19	Kot-19	19	30	12	22	12	26	00	00
20	Kot-20	21	10	00	20	24	30	02	05
21	Kot-21	12	00	00	00	14	18	00	00
22	Kot-22	10	22	14	20	18	18	00	16
23	Kot-23	10	14	14	12	20	24	12	00
24	Kot-24	12	22	12	18	18	24	22	16
25	Kot-25	00	20	08	12	16	20	00	24
26	Kot-26	16	12	20	18	18	28	12	20
27	Kot-27	00	25	26	00	20	19	00	00
28	Pseu-I	00	00	18	00	20	22	00	00
29	Pseu-II	00	00	00	00	20	22	00	00
30	Pseu-III	00	00	08	00	20	26	00	00

[Zone of inhibition is given in mm (millimeter)]

[Ch=Cephalothin, 30 µg; Cd=Clindamycin, 02 µg; Co=Co-Trimoxazole, 25 µg E=Erythromycin; 15 µg; G=Gentamycin, 10 µg; Of= Ofloxacin, 01 µg; P=Penicillin-G. 10 unit; Va=Vancomycin; 30 µg.]

TABLE: 3.9.2

ANTI-MICROBIAL ACTIVITY OF VARIOUS BACTERIAL ISOLATES

Sr.No.	Culture No.	Antibiotics Tested						
		Au	Ce	Tb	G	Pc	Ca	Cl
01	Kot-01	002	02	10	12	04	00	00
02	Kot-02	02	04	08	10	06	00	00
03	Kot-03	20	10	14	14	14	00	02
04	Kot-04	18	10	12	12	14	00	02
05	Kot-05	04	06	16	14	10	00	00
06	Kot-06	04	08	12	12	10	00	00
07	Kot-07	18	12	13	16	10	00	00
08	Kot-08	06	04	14	16	10	00	00
09	Kot-09	04	02	12	14	06	00	00
10	Kot-10	16	02	6	10	14	00	02
11	Kot-11	04	14	16	18	14	00	00
12	Kot-12	06	10	12	06	14	00	00
13	Kot-13	16	17	21	14	15	00	00
14	Kot-14	26	18	16	16	22	00	04
15	Kot-15	24	16	22	14	14	00	06
16	Kot-16	14	19	20	16	06	00	00
17	Kot-17	16	07	18	02	012	00	00
18	Kot-18	12	08	17	03	08	00	02
19	Kot-19	14	14	18	14	20	00	00
20	Kot-20	16	04	04	15	14	02	05
21	Kot-21	00	14	10	12	06	10	06
22	Kot-22	14	8	14	14	8	00	00
23	Kot-23	02	12	12	18	10	00	00
24	Kot-24	16	12	12	10	08	00	04
25	Kot-25	14	06	14	06	04	00	00
26	Kot-26	06	04	14	16	08	00	00
27	Kot-27	14	09	17	16	06	00	00
28	Pseu-I	00	06	12	14	12	12	04
29	Pseu-II	00	08	08	10	12	08	04
30	Pseu-III	00	08	16	14	14	00	04

[Zone of inhibition is given in mm (millimeter)]
[Au=Augmentin, 30 μg; Ce=Co-Cethotaxime, 10 μg; Tb=Tobramycin; 10 μg; G=Gentamycin, 10 μg; Pc= Piperacillin, 75 μg; Ca=Ceftazidime-G; 30 μg; Cl=Coistin; 25 μg.]

TABLE: 3.9.3

ANTI-MICROBIAL ACTIVITY OF VARIOUS BACTERIAL ISOLATES

Sr.No.	Culture No.	Antibiotics Tested						
		Ak	A	Ca	Ci	C	G	Pc
01	Kot-01	12	00	00	00	12	10	10
02	Kot-02	22	00	00	00	14	14	06
03	Kot-03	16	12	00	10	16	14	14
04	Kot-04	14	08	00	04	14	14	14
05	Kot-05	14	00	00	06	22	14	12
06	Kot-06	26	00	00	06	18	08	10
07	Kot-07	12	00	00	06	16	12	10
08	Kot-08	12	00	00	06	16	12	10
09	Kot-09	10	08	00	10	14	00	06
10	Kot-10	14	00	00	04	18	02	10
11	Kot-11	14	00	00	06	20	03	06
12	Kot-12	18	10	00	08	10	16	10
13	Kot-13	02	04	06	10	14	16	00
14	Kot-14	16	24	00	14	16	12	20
15	Kot-15	00	19	20	06	08	05	14
16	Kot-16	06	22	00	08	14	03	14
17	Kot-17	08	14	03	12	20	18	10
18	Kot-18	14	00	00	06	18	20	06
19	Kot-19	18	00	06	04	00	14	14
20	Kot-20	04	00	00	10	06	02	04
21	Kot-21	06	00	08	16	16	10	06
22	Kot-22	14	00	00	04	16	12	12
23	Kot-23	16	00	00	06	18	16	08
24	Kot-24	22	06	00	06	10	08	04
25	Kot-25	20	00	00	10	16	08	00
26	Kot-26	24	00	00	10	20	18	08
27	Kot-27	19	00	06	16	14	18	14
28	Pseu-I	10	00	02	00	00	8	06
29	Pseu-II	14	00	06	04	02	12	14
30	Pseu-III	14	00	14	08	06	12	16

[Zone of inhibition is in mm (millimeter)]

[Ak=Amikacin, 30 µg; A=Ampicillin, 10 µg; Cn=Cephoxitin, 30 µg; Ca=Ceftazidine; 30 µg; Ci=Ceftriaxone, 30 µg; C= Chloramphenicol, 30 µg; 10 mcg; Pc=Piperacillin; 100 µg.]

TABLE: 3.9.4

ANTI-MICROBIAL ACTIVITY OF VARIOUS BACTERIAL ISOLATES

Sr.No.	Culture No.	Antibiotics Tested					
		T	Cb	Cn	C	E	Mt
01	Kot-01	02	18	00	10	14	00
02	Kot-02	02	16	00	08	12	02
03	Kot-03	26	10	22	14	18	22
04	Kot-04	16	12	18	12	16	20
05	Kot-05	04	14	00	12	16	04
06	Kot-06	02	14	00	12	12	02
07	Kot-07	18	16	14	14	18	16
08	Kot-08	04	24	00	12	18	06
09	Kot-09	02	14	00	03	16	14
10	Kot-10	06	18	14	04	14	00
11	Kot-11	14	14	02	00	12	16
12	Kot-12	14	20	16	16	24	12
13	Kot-13	24	22	04	00	04	12
14	Kot-14	18	20	06	20	22	24
15	Kot-15	04	16	19	16	19	00
16	Kot-16	09	14	18	19	18	02
17	Kot-17	11	18	20	17	21	00
18	Kot-18	14	22	04	04	16	14
19	Kot-19	13	24	06	06	20	00
20	Kot-20	15	04	00	18	24	00
21	Kot-21	06	00	00	08	12	00
22	Kot-22	04	16	08	12	12	00
23	Kot-23	04	08	08	14	18	06
24	Kot-24	06	16	06	12	18	16
25	Kot-25	00	14	02	10	14	00
26	Kot-26	10	06	14	12	22	06
27	Kot-27	00	19	20	14	13	00
28	Pseu-I	00	00	12	14	16	00
29	Pseu-II	00	00	00	14	16	00
30	Pseu-III	00	00	02	14	20	00

[Zone of inhibition is given in mm (millimeter)]

[T=Tetracycline, 30 µg; Cb=Carbenicillin, 100 µg; Cn=Cephoxitin, 30 µg; C=Chloramphenicol, 30 µg; E= Erythromycin, 15 µg; Mt=Metronidazole; 05 µg]

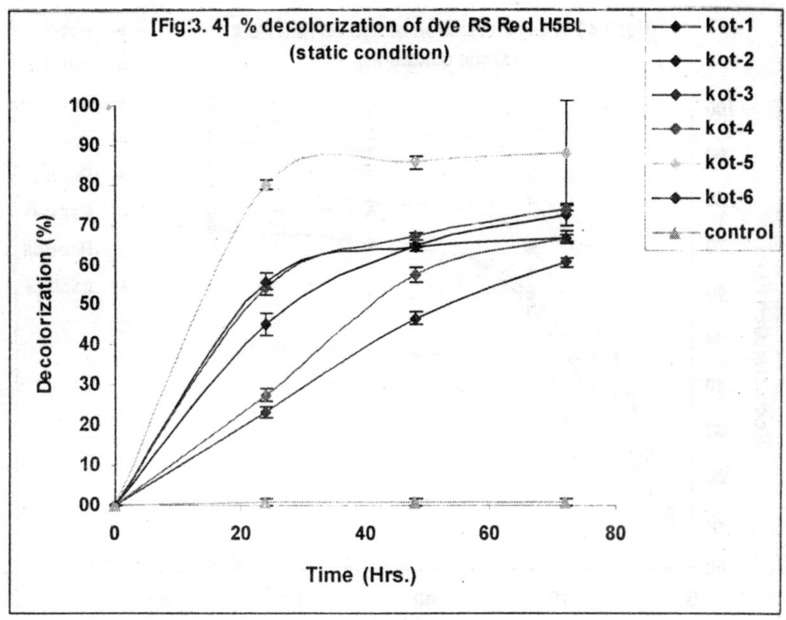
[Fig:3. 4] % decolorization of dye RS Red H5BL (static condition)
Decolorization (%)
100
90
80
70
60
50
40
30
20
10
00
0
20
40
60
80
Time (Hrs.)
kot-1
kot-2
kot-3
kot-4
kot-5
kot-6
control

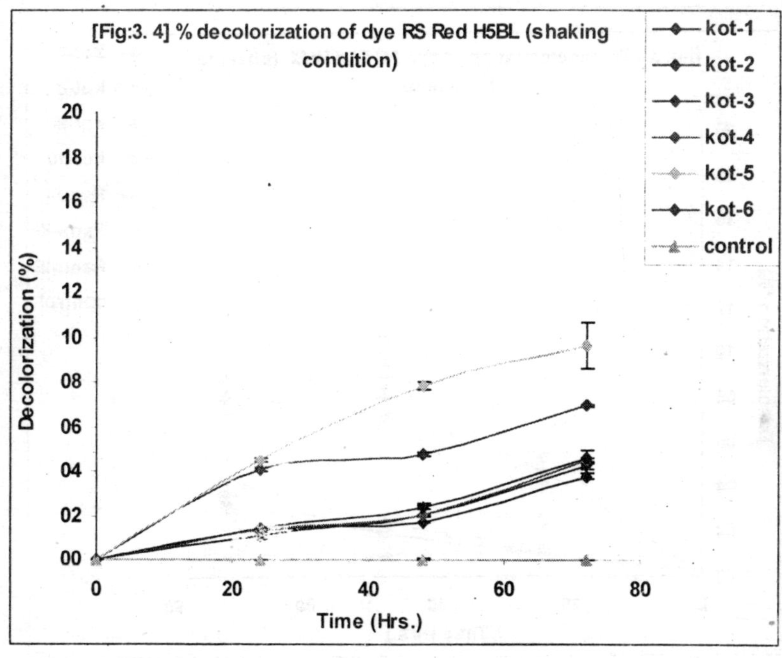
[Fig:3. 4] % decolorization of dye RS Red H5BL (shaking condition)
Decolorization (%)
20
18
16
14
12
10
08
06
04
02
00
0
20
40
60
80
Time (Hrs.)
kot-1
kot-2
kot-3
kot-4
kot-5
kot-6
control

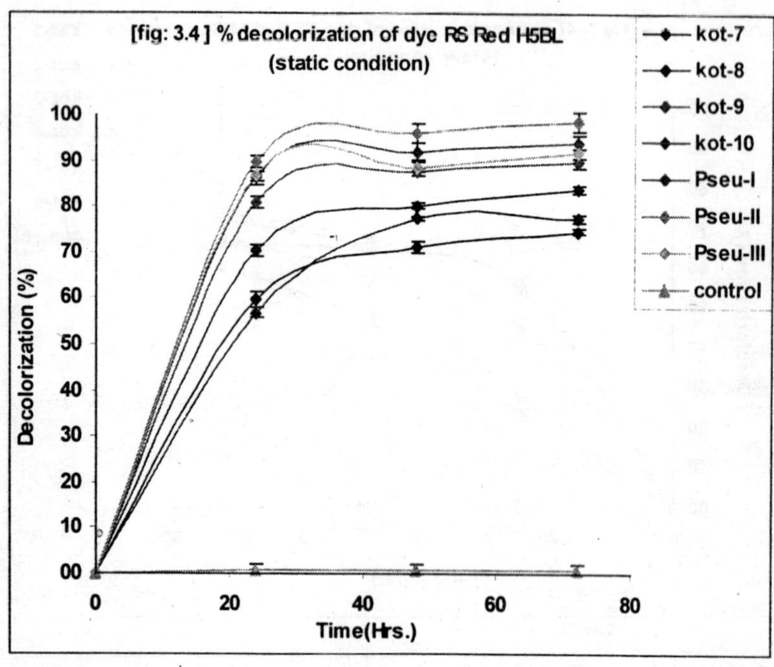
[fig: 3.4] % decolorization of dye RS Red H5BL
(static condition)
kot-7
kot-8
kot-9
kot-10
Pseu-I
Pseu-II
Pseu-III
control
Decolorization (%)
100
90
80
70
60
50
40
30
20
10
00
0
20
40
60
80
Time(Hrs.)

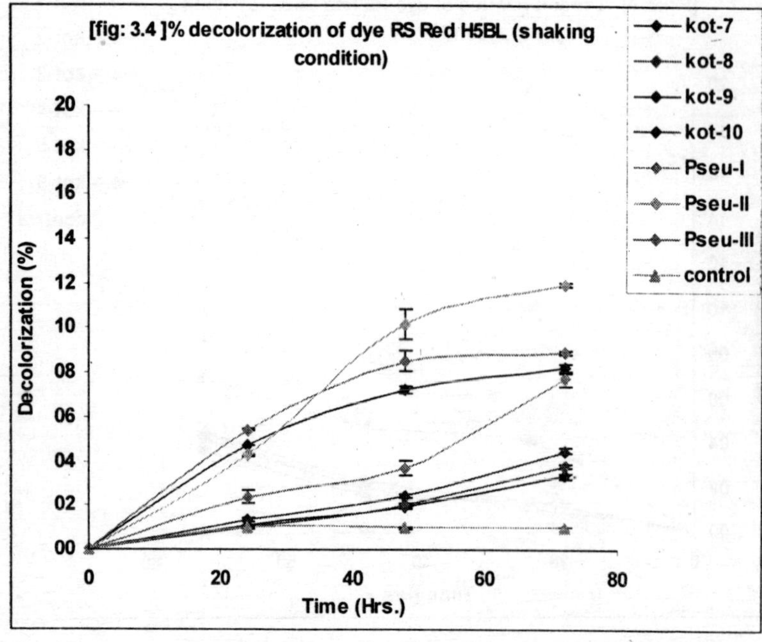
[fig: 3.4]% decolorization of dye RS Red H5BL (shaking
condition)
kot-7
kot-8
kot-9
kot-10
Pseu-I
Pseu-II
Pseu-III
control
Decolorization (%)
20
18
16
14
12
10
08
06
04
02
00
0
20
40
60
80
Time (Hrs.)

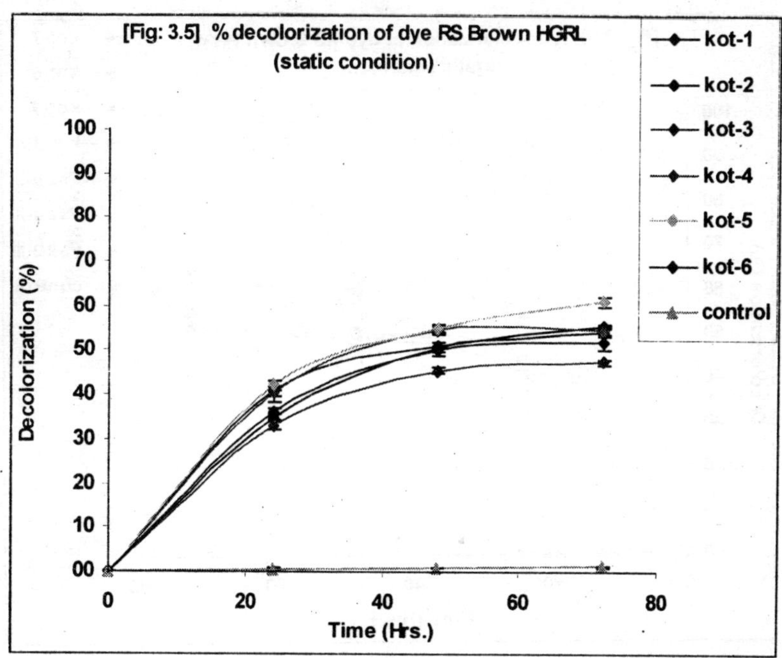
[Fig: 3.5] % decolorization of dye RS Brown HGRL
(static condition)
Decolorization (%)
100
90
80
70
60
50
40
30
20
10
00
0
20
40
60
80
Time (Hrs.)
kot-1
kot-2
kot-3
kot-4
kot-5
kot-6
control

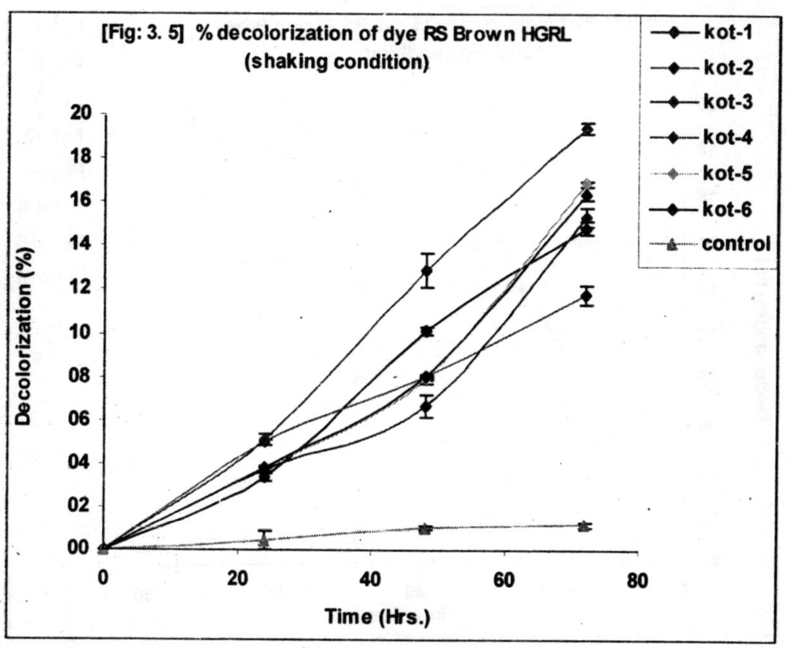
[Fig: 3. 5] % decolorization of dye RS Brown HGRL
(shaking condition)
Decolorization (%)
20
18
16
14
12
10
08
06
04
02
00
0
20
40
60
80
Time (Hrs.)
kot-1
kot-2
kot-3
kot-4
kot-5
kot-6
control

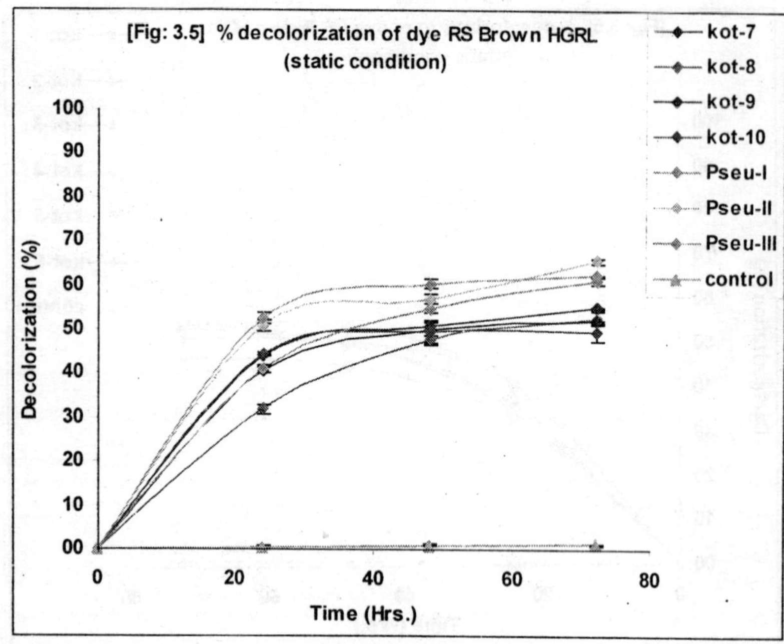
[Fig: 3.5] % decolorization of dye RS Brown HGRL
(static condition)
Decolorization (%)
100
90
80
70
60
50
40
30
20
10
00
0
20
40
60
80
Time (Hrs.)
kot-7
kot-8
kot-9
kot-10
Pseu-I
Pseu-II
Pseu-III
control

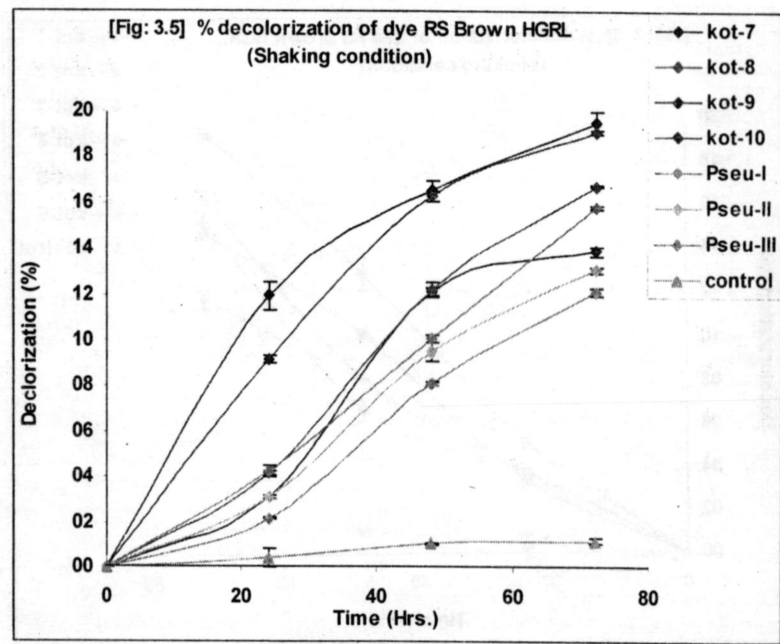
[Fig: 3.5] % decolorization of dye RS Brown HGRL
(Shaking condition)
Declorization (%)
20
18
16
14
12
10
08
06
04
02
00
0
20
40
60
80
Time (Hrs.)
kot-7
kot-8
kot-9
kot-10
Pseu-I
Pseu-II
Pseu-III
control

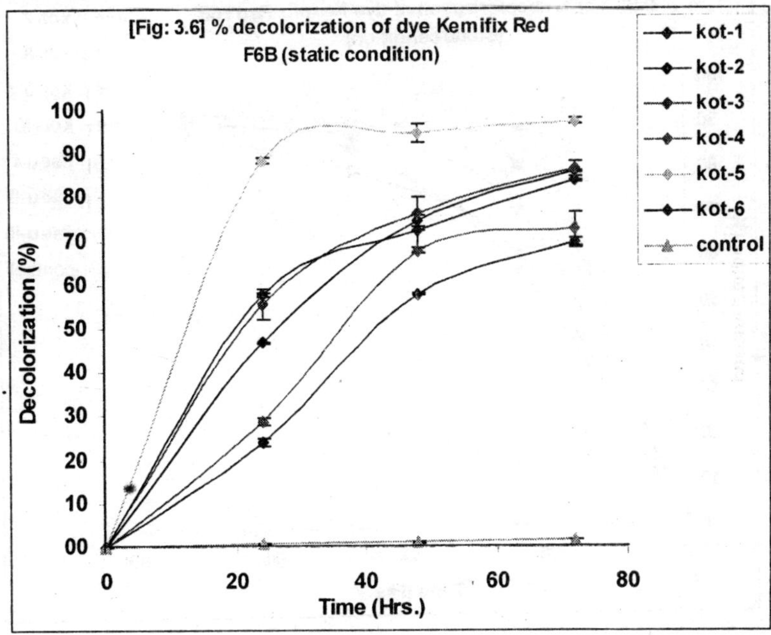
[Fig: 3.6] % decolorization of dye Kemifix Red F6B (static condition)
kot-1
kot-2
kot-3
kot-4
kot-5
kot-6
control
Decolorization (%)
100
90
80
70
60
50
40
30
20
10
00
0
20
40
60
80
Time (Hrs.)

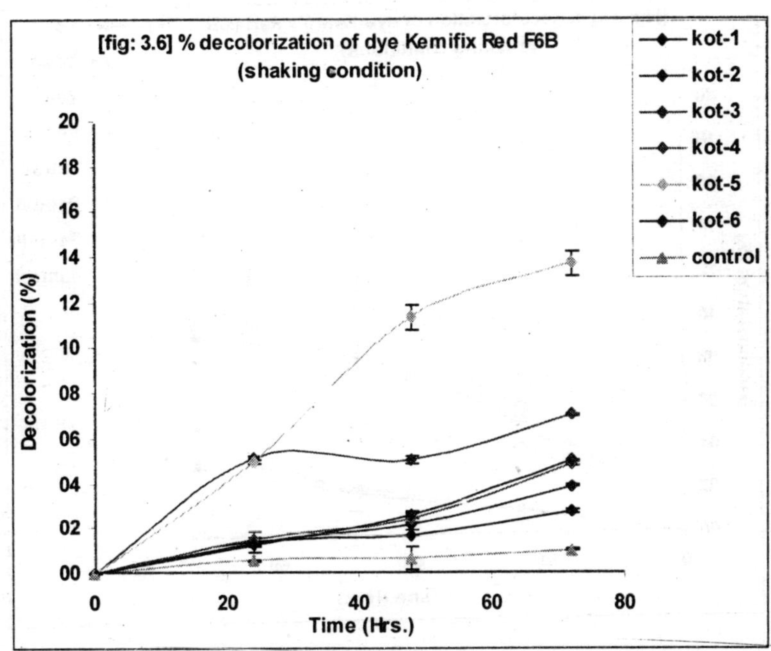
[fig: 3.6] % decolorization of dye Kemifix Red F6B (shaking condition)
kot-1
kot-2
kot-3
kot-4
kot-5
kot-6
control
Decolorization (%)
20
18
16
14
12
10
08
06
04
02
00
0
20
40
60
80
Time (Hrs.)

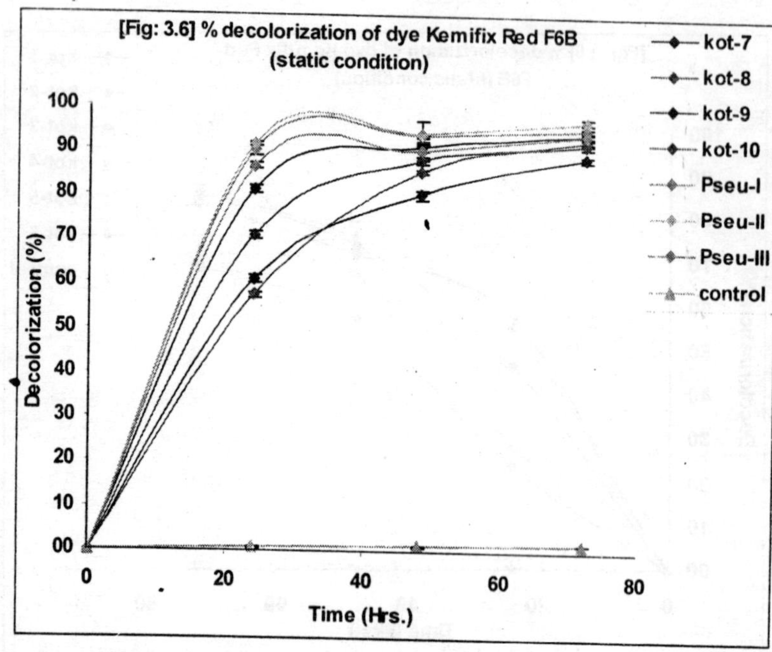
[Fig: 3.6] % decolorization of dye Kemifix Red F6B
(static condition)
kot-7
kot-8
kot-9
kot-10
Pseu-I
Pseu-II
Pseu-III
control
Decolorization (%)
100
90
80
70
60
50
40
30
20
10
00
0
20
40
60
80
Time (Hrs.)

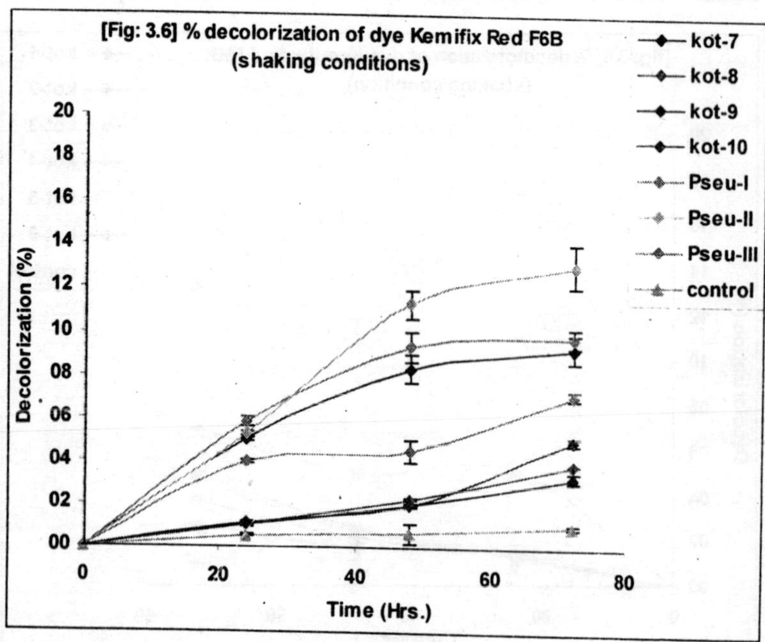
[Fig: 3.6] % decolorization of dye Kemifix Red F6B
(shaking conditions)
kot-7
kot-8
kot-9
kot-10
Pseu-I
Pseu-II
Pseu-III
control
Decolorization (%)
20
18
16
14
12
10
08
06
04
02
00
0
20
40
60
80
Time (Hrs.)

Dye RS Red H5BL was not effectively decolorized by the bacterial straine within 24 hours period except the strains kot-5, kot-9, Pseu-I, Pseu-II, and Pseu-III could decolorize 80% or more within 24 hours. It was also observed that the bacterial strains Pseu-I, Pseu-II, and Pseu-III had highest decolorization activity (above 90%) within 72 hours period. Under shaking condition, these strains showed a negligible reduction of color within 72 hours. The third dye (RS Brown HGRL) showed decolorization only between 47% to 65% within 72 hours under static condition.

The results clearly indicate that shake culture conditions did not favor decolorization dyes by any the strains. The controls were considered as the same except without inoculum.

The overall results show that the bacterial strains designated as Pseu-I, Pseu-II, and Pseu-III were the most effective and highest potential.

With respect to decolorization of individual dyes, added in the medium showed varied activity by each strain employed, *i.e.,* Kemifix Red F6B was very rapidly decolorized within 24 hours by all the three strains of *Pseudomonas* sp. The dye Red H5BL was also decolorized slightly less within 24 hours, whereas the third dye RS Brown HGRL showed 30 to 40% less decolorization than the other two dyes within 24 hours.

Thus, there was no significant decrease in the dye color, measured in shaking culture flasks as well as control.

EFFECTS OF COMBINATION(S) OF DIFFERENT NUTRIENT CONSTITUENTS ON DECOLORIZATION

Several bacterial strains were examined for their decolorization potentiality. In view of the previous experiments and results obtained, it was decided to study one of the most potential bacterial strain Pseu-II, in detail with respect to influence of certain constituents of nutrients and their combinations such as yeast extract, peptone, and benzoic acid along with the salts and glucose in the medium.

In the present study, the following combinations of nutrient ingredients were considered employing only one dye, *i.e.*, Red H5BL.

1. Glucose + KH_2PO_4 + $MgSO_4$,$7H_2O$ + Dye = [MGD]
2. Yeast extract+KH_2PO_4 + $MgSO_4$,$7H_2O$ + Dye = [MYED]
3. Peptone + KH_2PO_4 + $MgSO_4$,$7H_2O$ + Dye = [MPD]
4. Benzoic acid + KH_2PO_4 +$MgSO_4$, $7H_2O$ + Dye = [MBAD]
5. Glucose + Peptone + KH_2PO_4 + $MgSO_4$, $7H_2O$ + Dye = [G-1]
6. Glucose + Yeast extract + KH_2PO_4 + $MgSO_4$,$7H_2O$ + Dye = [G-2]
7. Yeast extract + Peptone + KH_2PO_4 + $MgSO_4$, $7H_2O$+ Dye = [G-3]
8. BA + Peptone + KH_2PO_4 + $MgSO_4$,$7H_2O$ + Dye = [P-1]
9. BA + Yeast extract + KH_2PO_4 + $MgSO_4$,$7H_2O$ + Dye = [P-2]

Under the stationary cultivation of Pseu-II in the medium, containing various combinations of nutrient sources, the decolorization assays were conducted at a regular interval of four hours and it was continued upto 60 hours. Assays were monitored by measuring decrease in absorbance and increase in biomass of the bacteria growth. Mean values of triplicate results were taken and presented graphically (Fig. 3.7, 3.8, and 3.9). Data were plotted after calculating the standard deviation) were taken

The values of document in absorbance and biomass under varying combinations were represented in (Fig: 3.7, 3.8 & 3.9).

1. When glucose was the only source of carbon and energy, the organisms showed maximum biomass that increased from 0.5 mg/ml to 1.64 mg/ml, while the absorbance value decreased from 0.66 to 0.56.
2. In presence of only yeast extract, there was increase in biomass (0.58 to 3.16 mg/ml) while the value of absorbance also decreased significantly (0.72 to 0.06).

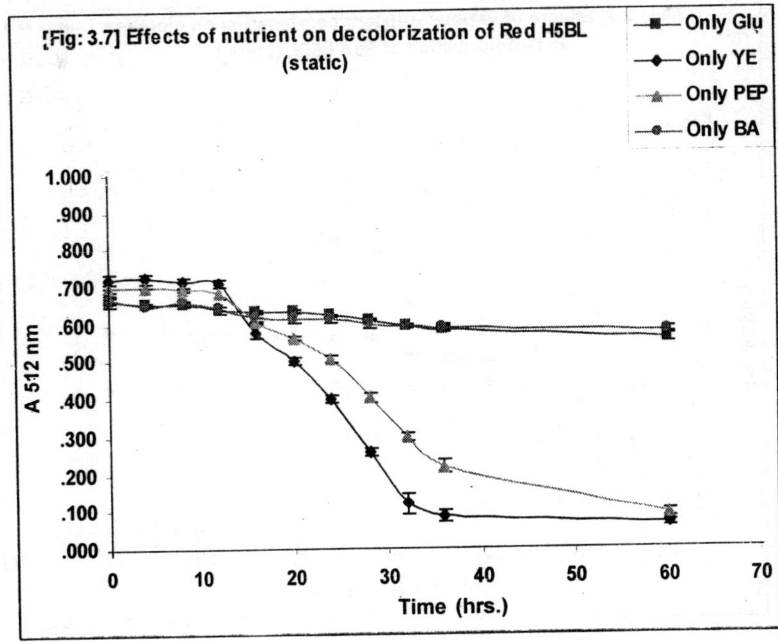

[Fig: 3.7] Effects of nutrient on decolorization of Red H5BL (static)
Only Glu
Only YE
Only PEP
Only BA
A 512 nm
1.000
.900
.800
.700
.600
.500
.400
.300
.200
.100
.000
0
10
20
30
40
50
60
70
Time (hrs.)

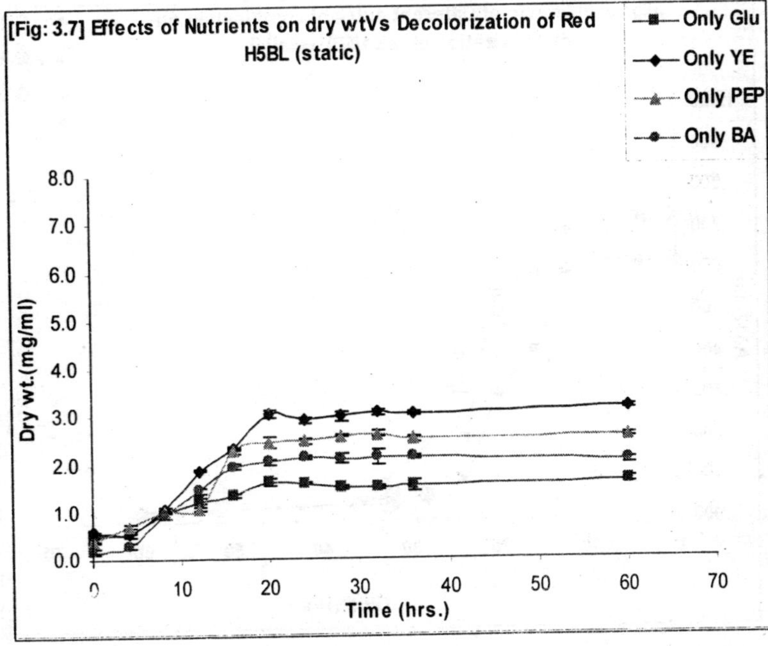

[Fig: 3.7] Effects of Nutrients on dry wtVs Decolorization of Red H5BL (static)
Only Glu
Only YE
Only PEP
Only BA
Dry wt.(mg/ml)
8.0
7.0
6.0
5.0
4.0
3.0
2.0
1.0
0.0
10
20
30
40
50
60
70
Time (hrs.)

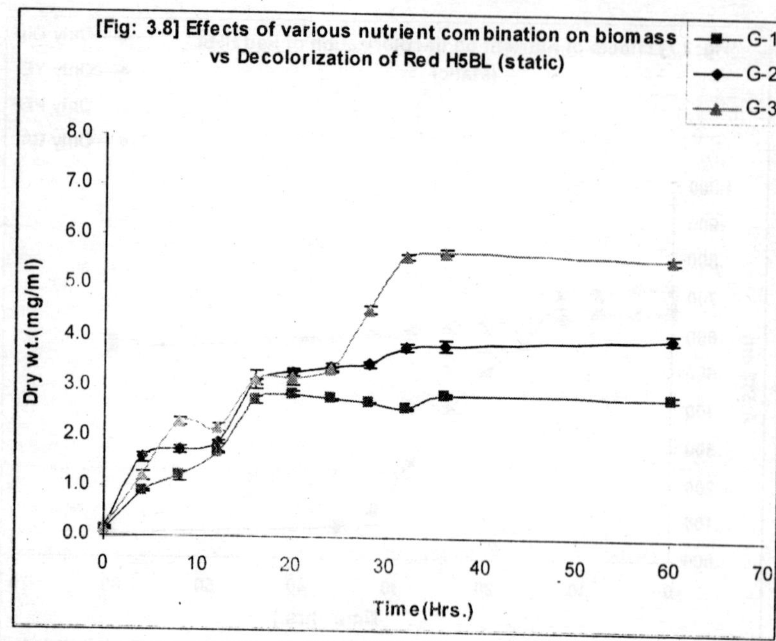
[Fig: 3.8] Effects of various nutrient combination on biomass vs Decolorization of Red H5BL (static)
G-1
G-2
G-3
Dry wt.(mg/ml)
8.0
7.0
6.0
5.0
4.0
3.0
2.0
1.0
0.0
0
10
20
30
40
50
60
70
Time(Hrs.)

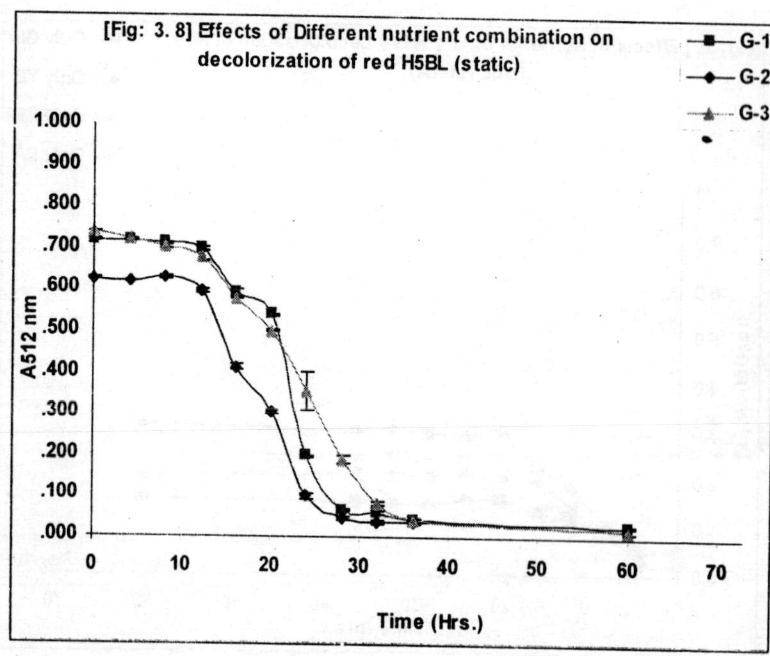
[Fig: 3. 8] Effects of Different nutrient combination on decolorization of red H5BL (static)
G-1
G-2
G-3
A512 nm
1.000
.900
.800
.700
.600
.500
.400
.300
.200
.100
.000
0
10
20
30
40
50
60
70
Time (Hrs.)

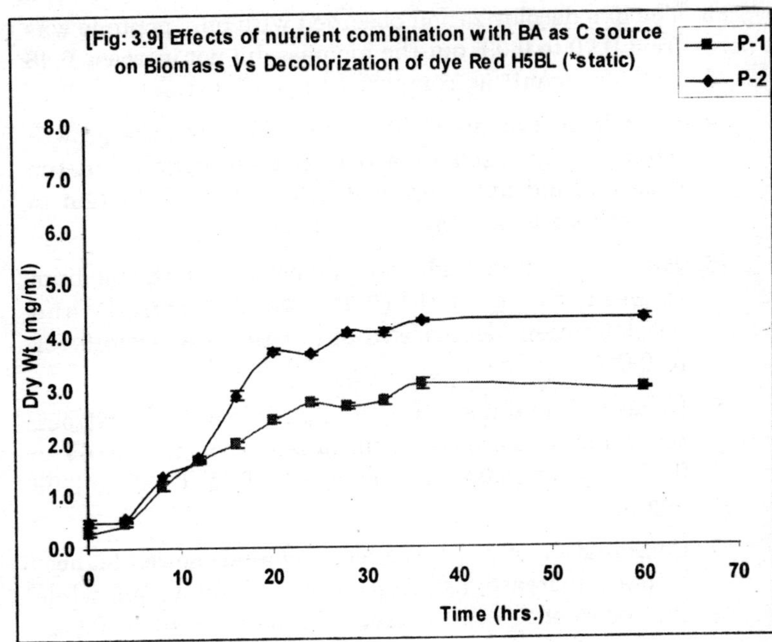
[Fig: 3.9] Effects of nutrient combination with BA as C source on Biomass Vs Decolorization of dye Red H5BL (*static)
P-1
P-2
Dry Wt (mg/ml)
8.0
7.0
6.0
5.0
4.0
3.0
2.0
1.0
0.0
0
10
20
30
40
50
60
70
Time (hrs.)

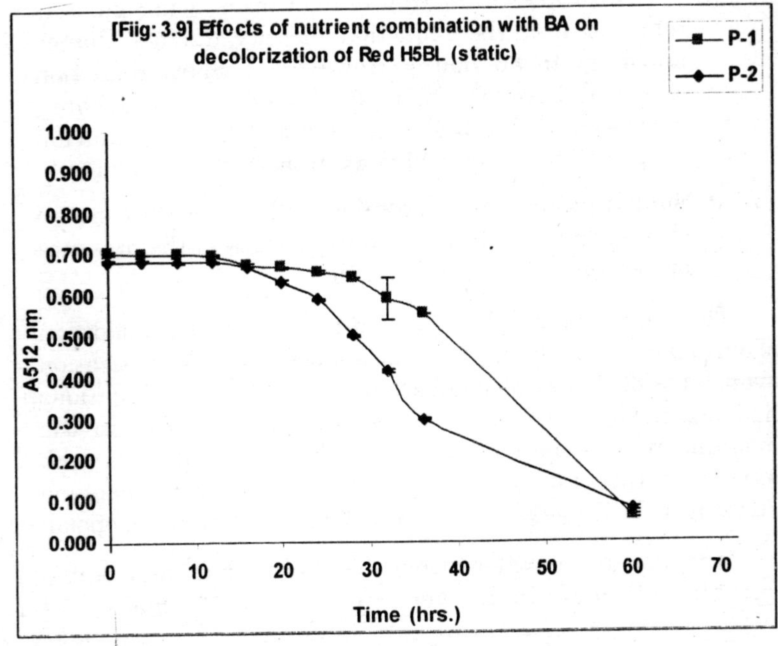
[Fiig: 3.9] Effocts of nutrient combination with BA on decolorization of Red H5BL (static)
P-1
P-2
A512 nm
1.000
0.900
0.800
0.700
0.600
0.500
0.400
0.300
0.200
0.100
0.000
0
10
20
30
40
50
60
70
Time (hrs.)

3. The dye decolorization observed with only peptone was from 0.70 to 0.09, but the biomass did not increase 0.48 to 2.58 mg/ml) as compared to yeast extract.
4. Benzoic acid as an carbon source favoured the growth (biomass increased from 0.47 to 2.05 mg/ml), but the condition did not support decolorization (reduction in absorbance was from 0.67 to 0.58).
5. The combination of glucose and peptone in the medium favored the growth (0.123 to 2.8 mg/ml) and decolorization (decrease in absorbance was from 0.722 to 0.046).
6. In case of combination having glucose + yeast extract, the value of decrease on absorbance was from 0.65 to 0.051 and increase in biomass was from 0.147 to 3.99 mg/ml.
7. Combination of yeast extract + peptone showed highest biomass increase, *i.e.*, from 0.143 to 5.68 mg/ml, while the absorbance value decreased from the 0.736 to 0.068.
8. The cultural medium containing benzoic acid + peptone reflected that the reduction in color did not change much up to 36 hours. However, a sharp reduction observed between 36 to 60 hours, *i.e.*, absorbance decreased from 0.553 to 0.064. The combination supported increase in biomass from 0.29 to 3.00 mg/ml).
9. Similar patterns were observed with respect to decrease in absorbance and increase in biomass in the presence of benzoic acid + yeast extract.

The results showed that the conditions in which the addition of only glucose or only benzoic acid, did not favour decolorization even after 60 hours of incubation, while other combinations had significance influence on decolorization of dye in the medium. With respect to increase in biomass, nutrients such as yeast extract, glucose + yeast extract, yeast extract + peptone and benzoic acid + yeast extract were found to be more beneficial.

It is also quite evident from the results that there was a notable relationship between increase in biomass and decolorization of dyes. It was also observed that after 8 hour of incubation, biomass approximately double in all the combinations, which then continued to increase upto 16 to 20 hours period, *i.e.,* almost the end of experimental phase. In almost all cases, rapid decrease in absorbance occurred after 24 to 28 hour period.

EFFECTS OF PERIODICAL LOADING OF DYES

Two culture flasks were incubated at 30°C, one containing only medium + inoculum + Dye and another flask containing medium + inoculum, for 12 hours under static conditions. After 12 hours, solution of dye RS Red H5BL and RS Brown HGRL added in the separate flasks, which was not containing dye, and further incubated. The flasks were harvested after every four hours interval upto 36 hours. Culture broths from both of flasks were centrifuged and supernatants were measured at the specific wavelengths for the remaining dye color.

The experiment was design to find out the incubation time at which the decolorization process was initiated, comparing with control. Results (Fig 3.10) clearly indicates that actual rates of decolorization were initiated after 12 hours in both the flasks, control and test flask in which dye was loaded at 12th hour. Thus, the organism required at least 12-hour period to achieve specific cell biomass and facultative conditions that might necessary to achieve dye decolorization process. The dye RS Brown HGRL was observed to be less susceptible compared to RS Red H5BL.

DOCUMENTATION OF BIODECOLORIZATION IN ASSAY MIXTURE

Using all the 20 dyes (0.02%) individually in the medium that inoculated (treated) and uninoculated (control or untreated) with the organism in test tubes were examined after 24 hours and decolorization of dyes recorded and presented in the photographs alongwith the control test tubes. For each dyes, three tubes (in triplicate) were taken and designated as, Cm: only medium; Cd: only dye solution; and T = medium + dye +

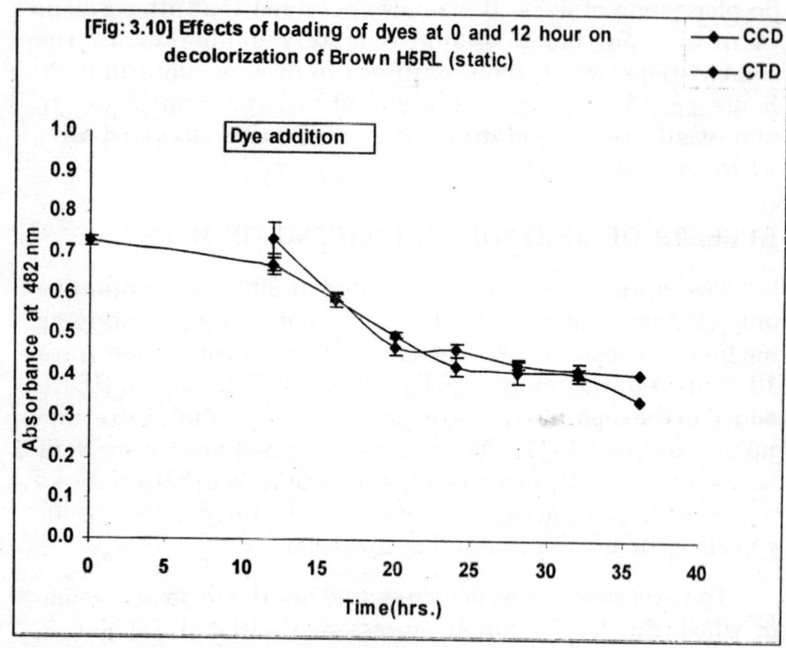
[Fig: 3.10] Effects of loading of dyes at 0 and 12 hour on decolorization of Brown H5RL (static)
CCD
CTD
Dye addition
Absorbance at 482 nm
1.0
0.9
0.8
0.7
0.6
0.5
0.4
0.3
0.2
0.1
0.0
0
5
10
15
20
25
30
35
40
Time(hrs.)

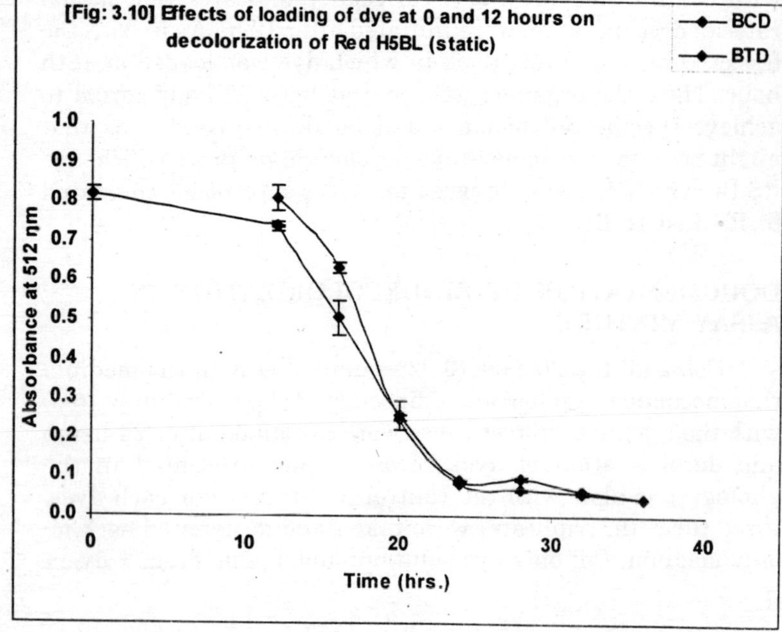
[Fig: 3.10] Effects of loading of dye at 0 and 12 hours on decolorization of Red H5BL (static)
BCD
BTD
Absorbance at 512 nm
1.0
0.9
0.8
0.7
0.6
0.5
0.4
0.3
0.2
0.1
0.0
0
10
20
30
40
Time (hrs.)

inoculum. The experiment performed to show that the organisms have potential to decolorize any textile dyes completely.

UV-Visible Spectrophotometric and Chromatographic Analysis of Decolorization of Dyes

The qualitative information can be useful in finding out the removal of color of dye(s) in the incubated assay mixture with and without inoculum, by monitoring on UV-Visible spectrophotometric and chromatographic analysis. The supernatants obtained at zero hour and after 72 hours of incubation, with inoculum, were scanned between 190 to 1100 nm.

UV-Visible Spectra of Assay Mixture

Decolorized sample of all the seven dyes selected were analyzed by scanning on UV-Visible Spectrophotometer (Fig. 3.11 to 3.17). Results of spectra for both the samples treated (medium + dye + inoculum) and untreated (medium + dye) were compared and interpreted as the change in absorbance value at the specific wavelength (λ_{max} nm) after inoculation of 24 hours.

Figures 3.11 to 3.17 show the result of changes in the spectra of the assay mixture is treated as well as control samples.

Figures 3.11 to 3.17 UV-Visible scans of dyes Kemifix Red F6B, RS Red H5BL, RS Violet H5RL, RS Navy Blue H2GL, RS Brown HGRL, Red H8B, and Red M5B, with their λ_{max}.

From zero to 72 hours all the dyes decolorized and reduced to above 90%. Disappearance of peaks at their specific λ_{max} indicate that the chromophore groups of all the seven dyes were removed and thus dyes completely decolorized after 72 hours of incubation. When all the scans shown in the figures were compared between the ranges of 190 to 373 nm wavelengths, there appeared to be formation of new bands (peaks) of various compounds. The spectra also indicate that the dyes were not only decolorized but might also be further metabolized.

MODE	: Spectrum	REF/SAMPLE	: Kemifix Red F6B
STARTING WL	: 190.0 (nm)	ENDING WL	: 1100.0 (nm)
WL INTERVAL	: 1.0 (nm)		
ANALYST	: Mr. Ramesh K. Kothari		
REMARKS	: MD= Medium+Dye, MDO=Medium+Dye+Org. (Treated)		

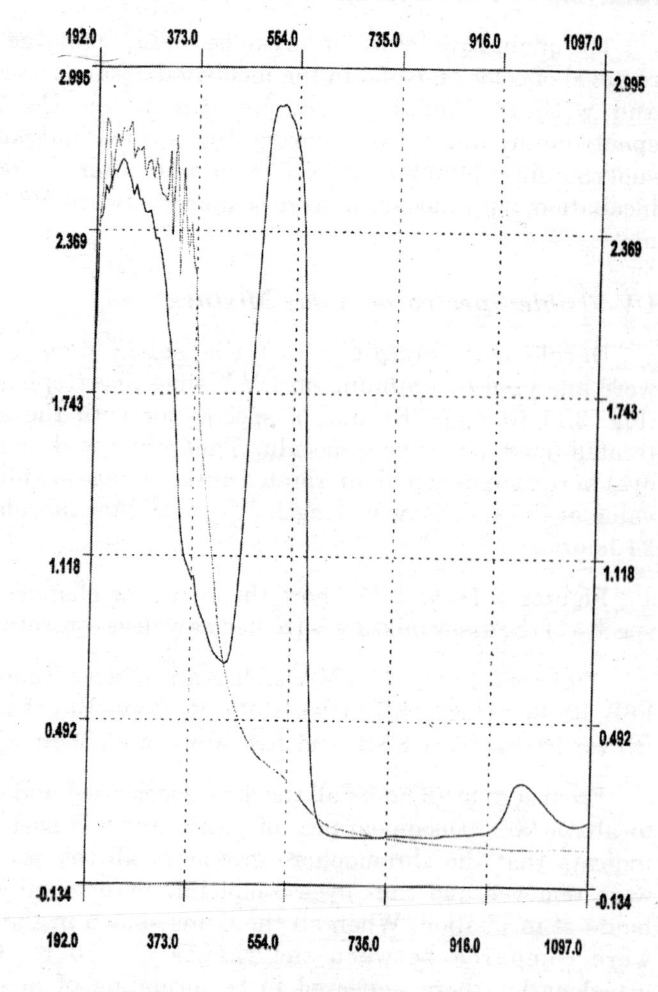

MODE	: Spectrum	REF/SAMPLE	:	RS Red H5BL
STARTING WL	: 190.0 (nm)	ENDING WL	:	1100.0 (nm)
WL INTERVAL	: 1.0 (nm)			
ANALYST	: Mr. Ramesh K. Kothari			
REMARKS	: MD= Medium+Dye, MDO=Medium+Dye+Org. (Treated)			

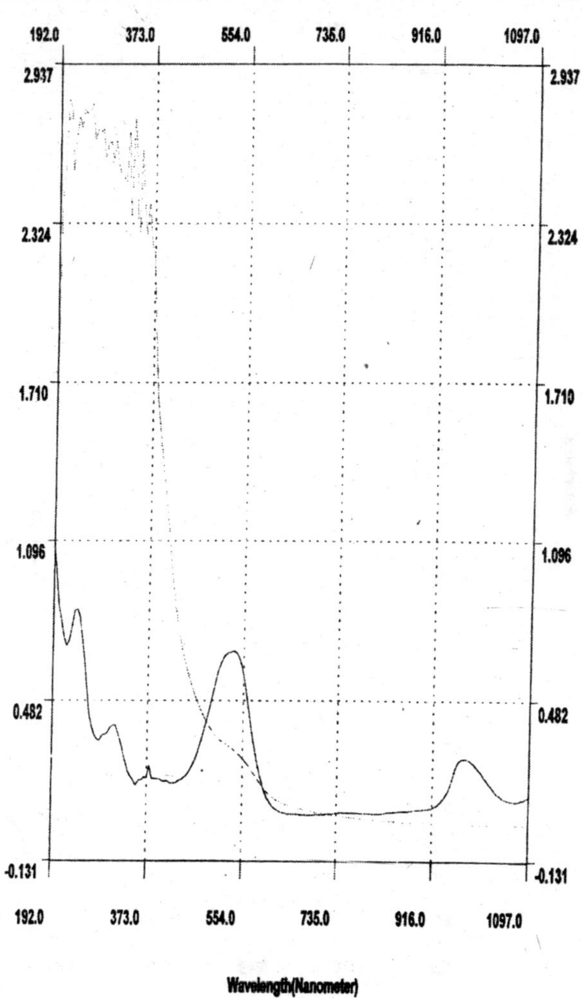

MODE	: Spectrum	REF/SAMPLE	: RS Violet H5RL
STARTING WL	: 190.0 (nm)	ENDING WL	: 1100.0 (nm)
WL INTERVAL	: 1.0 (nm)		
ANALYST	: Mr. Ramesh K. Kothari		
REMARKS	: MD= Medium+Dye, MDO=Medium+Dye+Org. (Treated)		

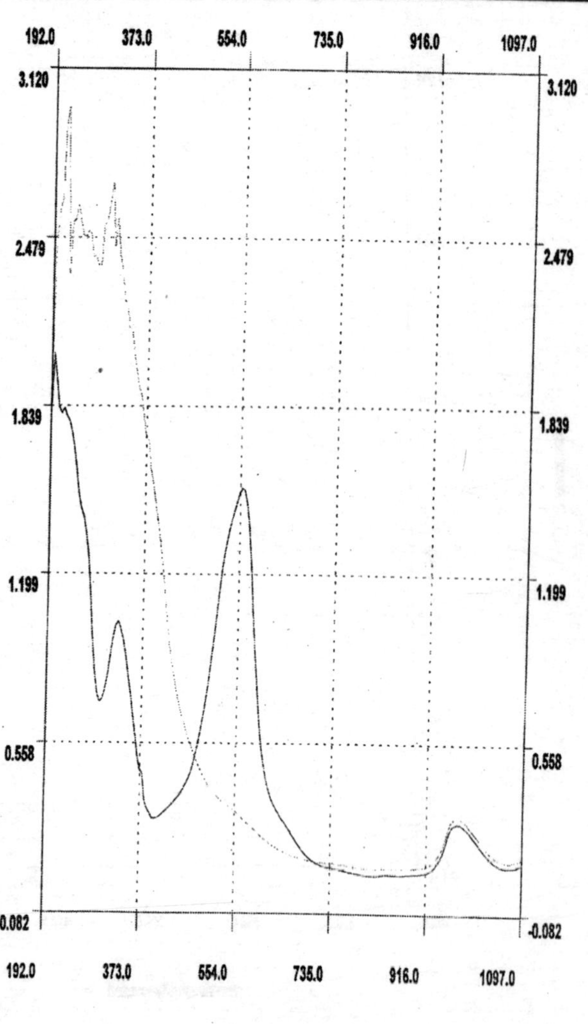

MODE : Spectrum REF/SAMPLE : RS Navy Blue H2GL
STARTING WL : 190.0 (nm) ENDING WL : 1100.0 (nm)
WL INTERVAL : 1.0 (nm)
ANALYST : Mr. Ramesh K. Kothari
REMARKS : MD= Medium+Dye, MDO=Medium+Dye+Org. (Treated)

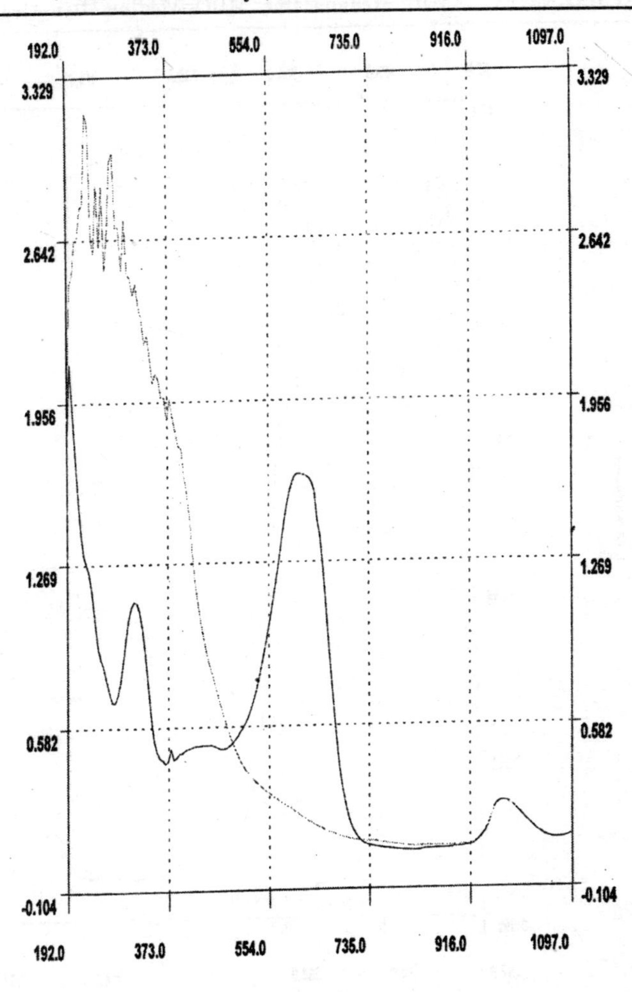

MODE	: Spectrum	REF/SAMPLE	: RS Brown HGRL
STARTING WL	: 190.0 (nm)	ENDING WL	: 1100.0 (nm)
WL INTERVAL	: 1.0 (nm)		
ANALYST	: Mr. Ramesh K. Kothari		
REMARKS	: MD= Medium+Dye, MDO=Medium+Dye+Org. (Treated)		

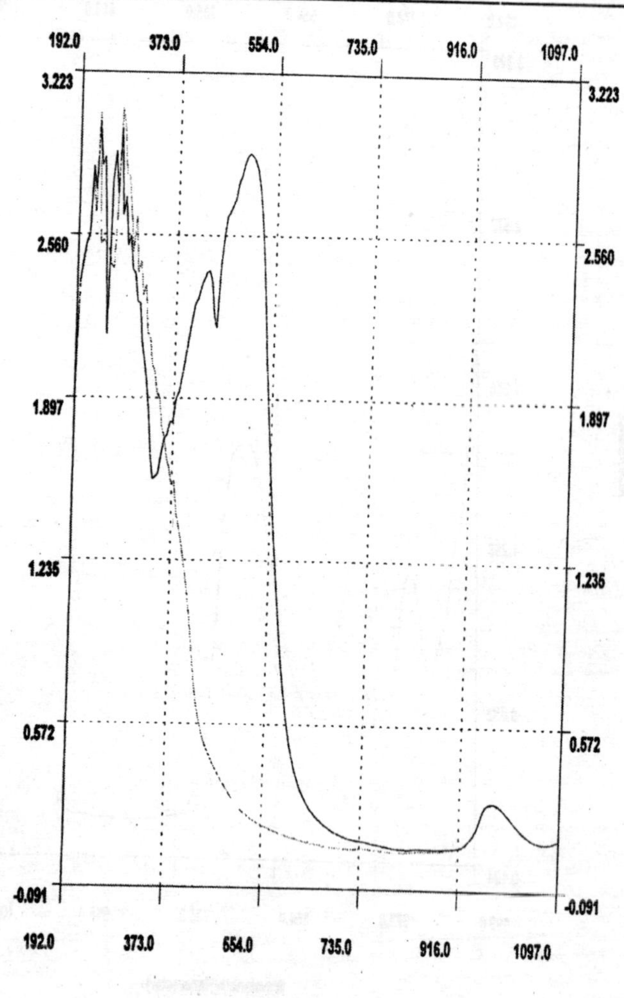

MODE	:	Spectrum	REF/SAMPLE	:	Red H8B
STARTING WL	:	190.0 (nm)	ENDING WL	:	1100.0 (nm)
WL INTERVAL	:	1.0 (nm)			
ANALYST	:	Mr. Ramesh K. Kothari			
REMARKS	:	MD= Medium+Dye, MDO=Medium+Dye+Org. (Treated)			

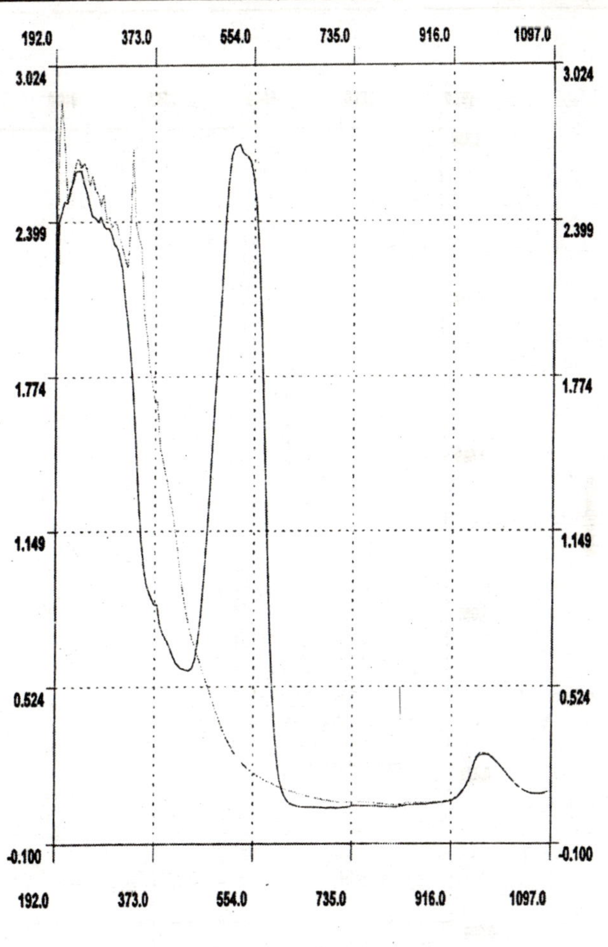

MODE	:	**Spectrum**	**REF/SAMPLE**	:	**Red H5B**
STARTING WL	:	**190.0 (nm)**	**ENDING WL**	:	**1100.0 (nm)**
WL INTERVAL	:	**1.0 (nm)**			
ANALYST	:	**Mr. Ramesh K. Kothari**			
REMARKS	:	**MD= Medium+Dye, MDO=Medium+Dye+Org. (Treated)**			

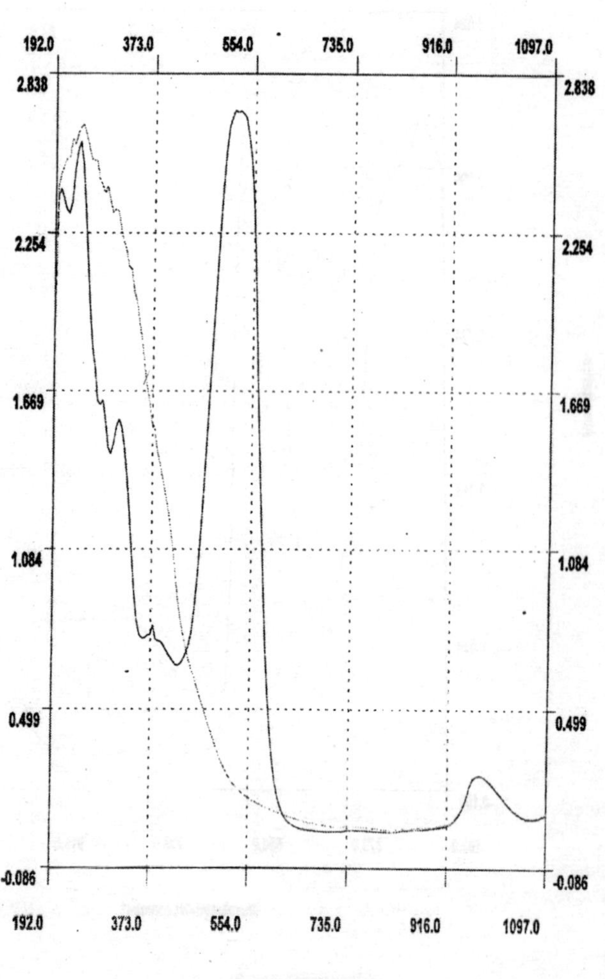

Evalution of Chromatogram

Description of the Samples

Four samples were prepared and used for chromatographic run on TLC plates:

1. Dye solution in methanol.
2. Supernatant of only medium (zero hours).
3. Supernatant of medium + inoculum (72 hours of incubation).
4. Supernatant of medium + inoculum + dye (72 hours of incubation).

These samples were run using two mobile phase, methano + chloroform, and methanol + ethyl acetate after spotting the four samples on the separate TLC plates (Plate 3.1 to 3.20). Observation was made under illumination at 254 nm and 365 nm. For each dye sample, the Rf values were calculated and (Table.3.10). The sample 2, 3 and 4 were observed on the plates showed that the sample 2 contained two components under the short and far UV illumination for all the 20 assay mixtures, while the sample 3 indicated two fractions separated on the plates and in the sample 4, the dye fraction did not appear but the medium fraction was present.

HPLC and GC Profile Studies

In this study, various aromatic amines were selected, since these amines are the chemical constituents of the several textile dyes. Two aromatic amines were added in the medium with inoculum [here cells bacterial strain Pseu-II were used as inoculum],and incubated under highly aerobic conditions. The sample were analyzed by HPLC and GC after incubation, are shown in (Fig 3.18 and 3.19).

Before and after incubation samples subjected to HPLC and GC analysis indicated that the two aromatic amines were degraded by comparing the changes in the sizes (areas) of peaks noted at 16.00 and 21.00 minutes and 8.0 min, 9.8 min, and 22.00 min peaks.

TABLE: 3.10

THIN LAYER CHROMATOGRAPHY: RF VALUES AND SOLVENT SYSTEMS OF SELECTED TEXTILE DYES

Sr.No	Dye Code	Name of Dye	Solvent System	Rf Value
1	Jd-00	Kemifix Red F6B	Methanol: Chloroform	0.62
2	Jd-01	RS Red H5BL	Methanol: Chloroform	0.91
3	Jd-02	RS Violet H5RL	Methanol: Chloroform	0.64
4	Jd-03	RS Navy Blue H2GL	Methanol: Chloroform	0.92
5	Jd-04	RS Brown HGRL	Methanol: Chloroform	0.56
6	Jd-05	Red H8B	Methanol: Chloroform	0.80
7	Jd-06	Red M5B	Methanol: Ethyl acetate	0.64
8	Jd-07	Purple H3R	Methanol: Ethyl acetate	0.62
9	Jd-08	Orange 3R	Methanol: Chloroform	0.90
10	Jd-09	Reactive Blue-81	Methanol: Chloroform	0.78
11	Jd-10	Acid Red-249	Methanol: Chloroform	0.91
12	Ad-11	Pc-Black-HN	Methanol: Ethyl acetate	0.80
13	Ad-12	Red-HB	Methanol: Chloroform	0.88
14	Ad-13	T-Blue-G	Methanol: Ethyl acetate	0.80
15	Ad-14	Yellow-FGM	Methanol: Chloroform	0.58
16	Ad-15	Red 6 Bx	Methanol: Ethyl acetate	0.87
17	Ad-16	Reactive Yellow-4	Methanol: Ethyl acetate	0.90
18	Kd-17	Red Brown HoR	Methanol: Chloroform	0.84
19	Kd-18	Golden-HR	Methanol: Ethyl acetate	0.65
20	Kd-19	Magenta-HB	Methanol: Chloroform	0.50

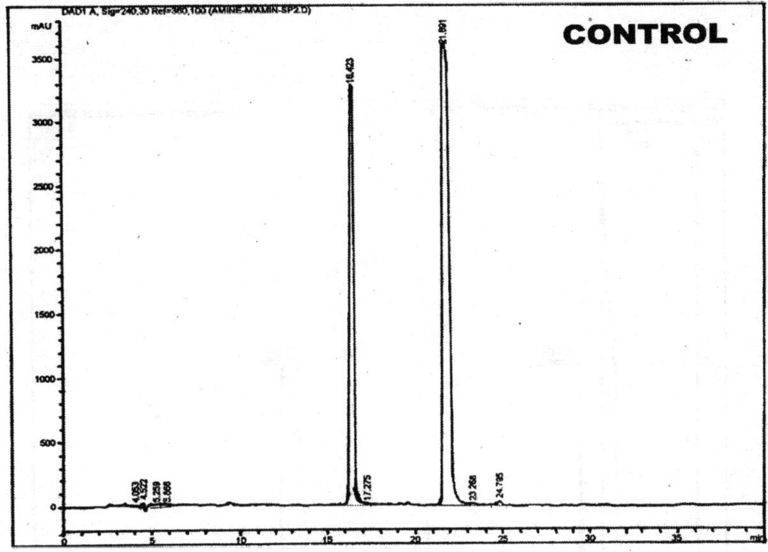

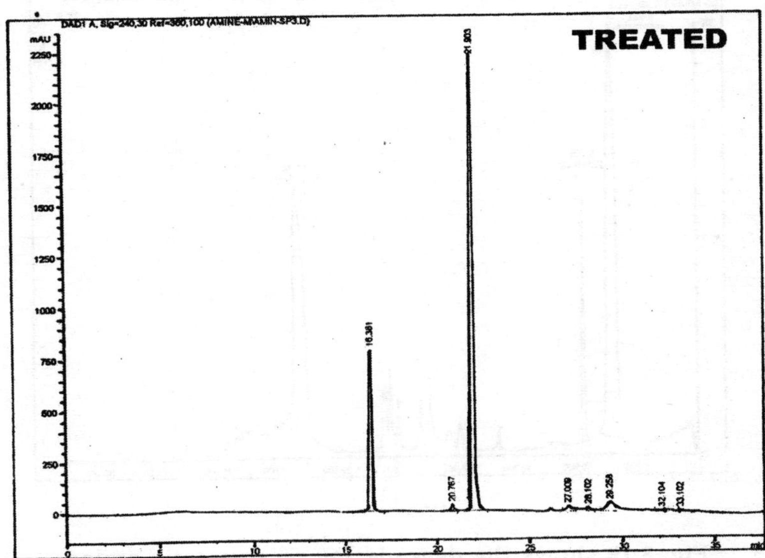

Fig. 3.18 shows the degradation of aeromatic amines after 24 hours of incubation.

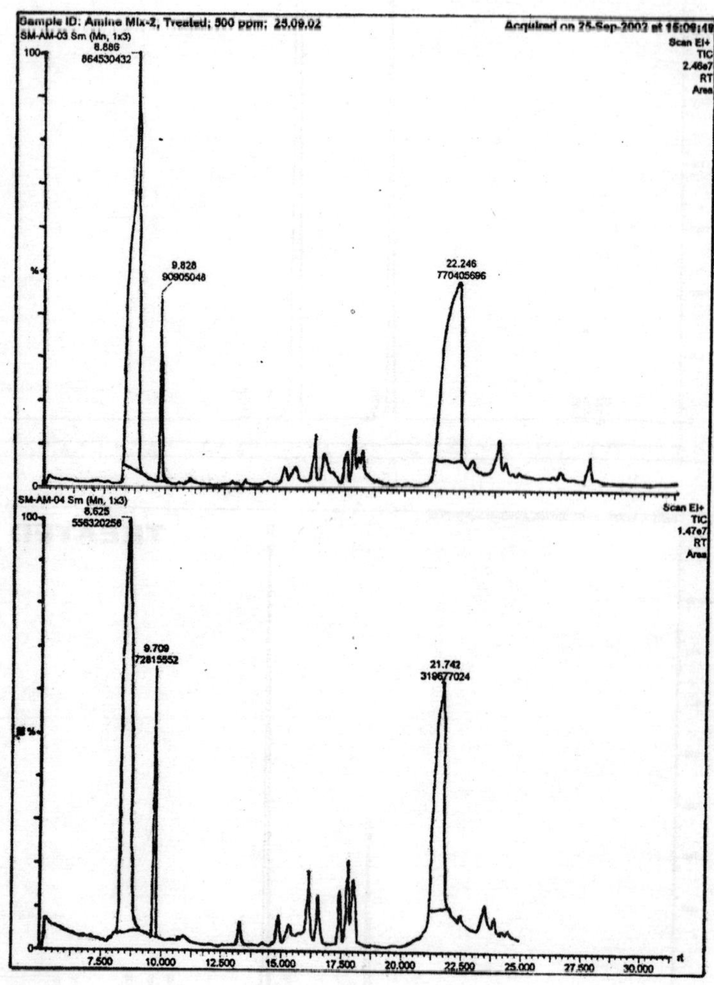

Fig. 3.19 shows the degradation of aeromatic amines after 24 hours of incubation.

INDUCTION OF RING CLEAVAGE BY *PSEUDOMONAS Sp.* STRAIN-II IN PRESENCE OF THE DYES

Five dyes were added in the medium to test induction of enzyme system responsible for the ring cleavage in the organisms. Results recorded for *ortho* and *meta* cleavages by the organism in presence of five dyes. In presence of dyes the organism showed induction of enzymes responsible for only *meta* cleavage process while the *ortho* cleavage process, there was not any positive result. The assay mixtures of all controls, *i.e.,* cultivation without dye in the medium, did not show any change in color indicating that the presence of dye(s) in the medium were required to induce the aromatic ring cleavage enzyme system. The following table shows the results recorded:

Dye Code	Name of Dyes	Pseudomonas sp. strain-II Ortho cleavage	Meta cleavage
Control	Without dyes	-	-
Jd-00	Kemifix Red F6B	-	+
Jd-01	RS Red H5BL	-	+
Jd-03	RS Violet H5RL	-	+
Jd-04	RS Navy Blue H2GL	-	+
Jd-05	RS Brown HGRL	-	+

GROWTH RESPONSE OF BACTERIAL ISOLATES AND RING CLEAVAGE OF PAHs

All the bacterial isolates, isolated and tentatively identified as *Pseudomonas sp.* Strain-I, (Pseu-I), *Pseudomonas sp.* Strain-II, (Pseu-II), and *Pseudomonas sp.* Strain-III (Pseu-III) were not able to utilize PAHs like, naphthalene, anthracene, phenanthrene, pyrene, chrysene, Benz[a]anthracene, and benzo[a]pyrene, fluoranthene, and acenaphthene in Minimal Medium (MM) as a sole source of carbon but when grown in a MM containing glucose as cosubstrate, they were showing ring cleavage activity. However, there were distinct differences between the strains with respect to kind of aromatic compound utilized and the rate of growth. Table 3.15 and 3.16 shows results of the batch experiments conducted with various bacterial isolates on different PAHs. Pseu-I was able to grow in MM containing naphthalene, anthracene, phenanthrene and pyrene (MM with glucose) and which shows the *meta* cleavage indicating ring opening. Besides the growth on four PAHs, and *meta* cleavage, Pseu-II was able to grow in MM containing, chrysene, Benz[a]anthracene, and benzo[a]pyrene (MM with glucose) also, giving *meta* cleavage of ring structures. Additionally, Pseu-III was able to grow in MM containing, naphthalene, anthracene, phenanthrene, pyrene, and chrysene (with glucose), which showing ortho cleavage action on the ring.

WHITE ROT FUNGI AND DYES

Dye decolorization by concentrated crude extracellular enzyme prepared from *Coriolopsis polysona* and newly isolated white rot fungus species Strain Rc infested wheat straw.

Many white rot fungi nowadays have got their place in remediation of environment polluted with nondegradable pollutants. For *e.g., Tremetus versicolor* produces various extracellular ligninolytic enzymes during its growth on wheat straw, Vyas *et al.* (1994a)

In the present work, the production of various extracellular ligninolytic enzymes (MnP, MIP, and Laccase) were studied and their applications in the decolorization of various synthetic textile dyes like, Kemifix Red F6B, RS Red H5BL, and RS Brown HGRL was studies.

Extracellular fluid of *Coriolopsis polysona* grown on wheat straw medium when used as the source of ligninolytic enzymes decolourised all the dyes tested but varying extent (Fig 3.20.1, 3.20.2, 3.21.1, 3.21.2, 3.22.1, and 3.22.2).

The solid-state fermentation technique was employed to generate the enzyme systems. Chopped wheat-straw was used to grow the fungus *C. polysona* and activity of the enzymes were estimated from day 3 to 21. The maximum activity of these 3 enzymes were observed on the day 14.

TABLE 3.11 : RING CLEAVAGE OF VARIOUS POLYCYCLIC AROMATIC HYDROCARBONS BY THREE STRAINS OF *PSEUDOMONAS* (PSEU-I, PSEU-I, ND PSEU-III).

Sr. No	Name of PAHs	Pseu-I		Pseu-II		Pseu-III	
		Ortho	Meta	Ortho	Meta	Ortho	Meta
1.	Naphthalene	-	+	-	+	+	-
2.	Anthracene	-	+	-	+	+	-
3.	Phenanthrene	-	+	-	+	+	-
4.	Pyrene	-	+	-	+	+	-
5.	Chrysene	-	-	-	+	+	-
6.	Benz[a]-anthracene	-	-	-	+	-	-
7.	Benzo[a]-pyrene	-	-	-	+	-	-
8.	Fluoranthene	-	-	-	-	-	-
9.	Acenaphthene	-	-	-	-	-	-

[+ indicate positive ring cleavage reaction—indicate negative ring cleavage reaction].

TABLE 3.12 : GROWTH RESPONSE OF THREE STRAINS OF *PSEUDOMONAS* (PSEU-I, PSEU-II, PSEU-III) ON VARIOUS POLYCYCLIC AROMATIC HYDROCARBONS

Sr. No	Name of PAHs	Growth Response		
		Pseu-I	Pseu-II	Pseu-III
1.	Naphthalene	+++	+++	+++
2.	Anthracene	+++	+++	++
3.	Phenanthrene	++	+++	++
4.	Pyrene	+	++	+
5.	Chrysene	-	+	+
6.	Benz[a]-anthracene	-	+	-
7.	Benzo[a]-pyrene	-	+	-
8.	Fluoranthene	-	-	-
9.	Acenaphthene	-	-	-

[+ indicate positive ring cleavage reaction, –indicate negative ring cleavage reaction].

Desalted extracellular enzymes prepared from *C. polysona* and strain *Rc* grown on wheat-straw, have been experimented to their decolorizing action on three dyes Kemifix Red F6B, RS Red H5BL, and RS Brown HGRL. Three enzymes tested were Mn–dependent peroxidase (Mn), Mn–independent peroxidase (MIP) and Laccase. The dyes have been used as substrates for these three enzymes MnP ad MIP.

The rate of decolorization of selected dyes by MnP, MIP and Laccase enzymes were expressed as the mean difference in absorbance during the time of incubation of RM (0.0-5.0 min).

The values of decolorization obtained for MnP, MIP, and Laccase by *C. polysona* and strain Rc are shown in Figure 3.20.1, 3.20.2. 3.21.1, 3.21.2, 3.22.1, and 3.22.2. The results showed that MnP activity was quite effective on dye Kemifix Red F6B highest decolorization (A 1.230, by Rc, and A 0.933, by Cp) at pH 4. MnP activity on dyes RS Red H5BL show noticeable decolorization (A 0.372, by Rc and A 0.319) at pH 4. MnP activity on dye RS Brown HGRL show also considerable decrease in absorbance (A 0.300, by Rc and A 0.241, by Cp) at pH 4.

MIP activity show comparatively less decolorization of dyes Kemifix Red F6B, RS Red H5BL, and RS Brown HGRL (A 0.301, by Rc, and A 0.229, by Cp, A 0.086, by Rc, and A 0.066, by Cp, and A 0.131, by Rc and A 0.103, by Cp respectively, at pH 3.5, than the MnP activity.

Decolorization of dyes by Laccase activity was very less compared to MnP, and MIP activity. Dyes Kemifix Red F6B, RS Red H5BL, and RS Brown HGRL shows very less decrease in absorbance (A-0.110, by Rc, and A-088, by Cp, A-0.138, by Rc, and A-0.110, by Cp, and A-0.136, by Rc and A-0.098, by Cp respectively, at pH 4.0).

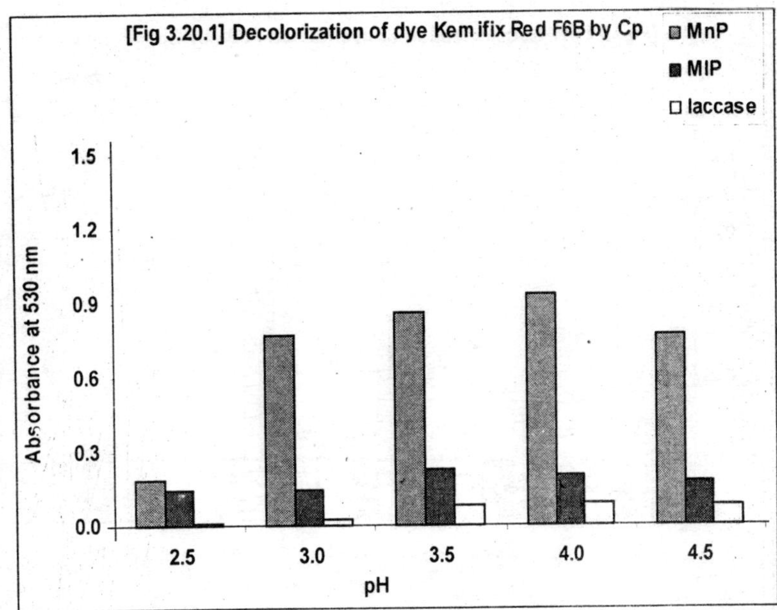
[Fig 3.20.1] Decolorization of dye Kemifix Red F6B by Cp
MnP
MIP
laccase
1.5
1.2
0.9
0.6
0.3
0.0
Absorbance at 530 nm
2.5
3.0
3.5
4.0
4.5
pH

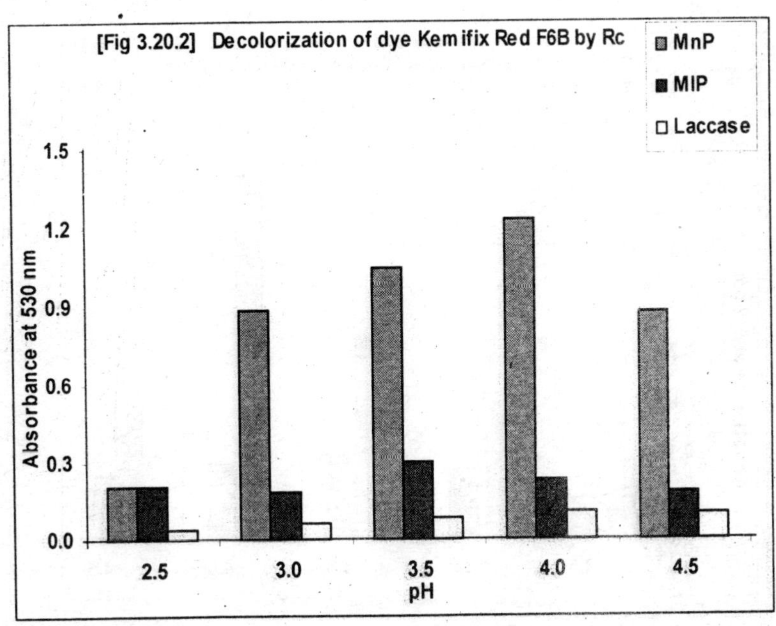
[Fig 3.20.2] Decolorization of dye Kemifix Red F6B by Rc
MnP
MIP
Laccase
1.5
1.2
0.9
0.6
0.3
0.0
Absorbance at 530 nm
2.5
3.0
3.5
4.0
4.5
pH

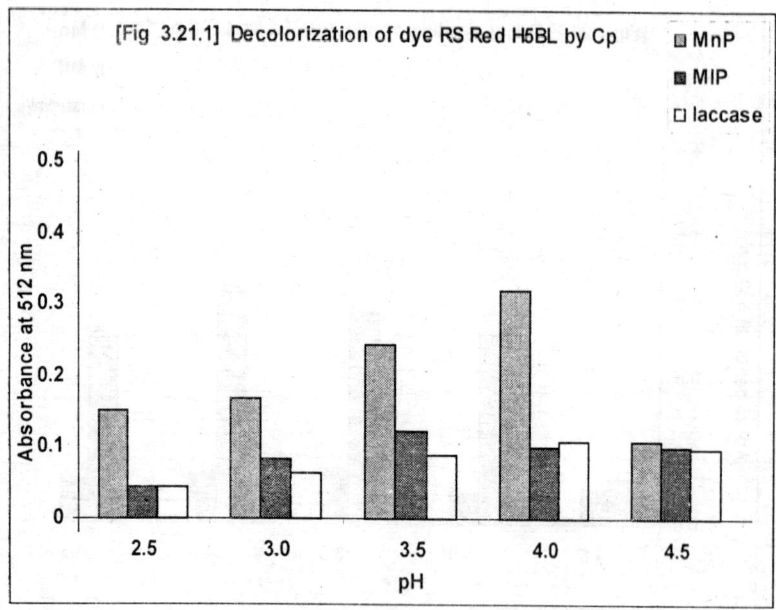
[Fig 3.21.1] Decolorization of dye RS Red H5BL by Cp
MnP
MIP
laccase
Absorbance at 512 nm
0.5
0.4
0.3
0.2
0.1
0
2.5
3.0
3.5
4.0
4.5
pH

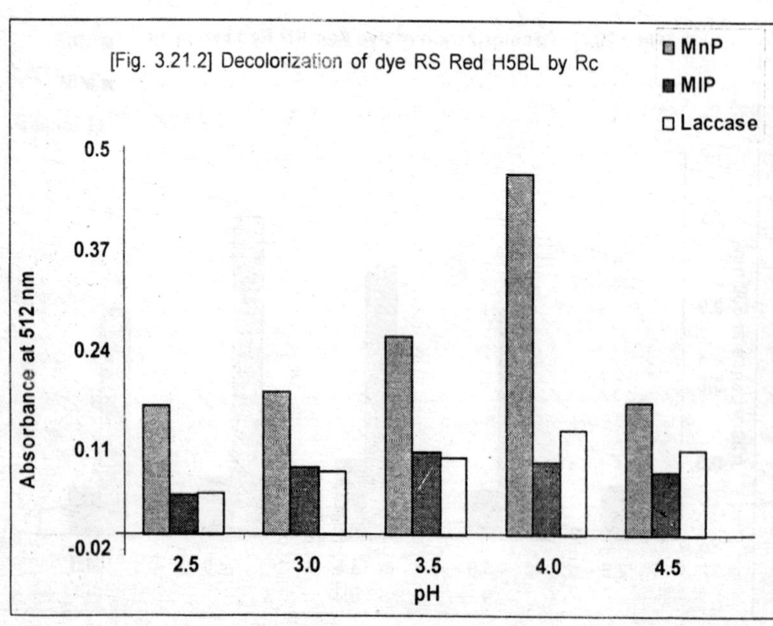
[Fig. 3.21.2] Decolorization of dye RS Red H5BL by Rc
MnP
MIP
Laccase
Absorbance at 512 nm
0.5
0.37
0.24
0.11
-0.02
2.5
3.0
3.5
4.0
4.5
pH

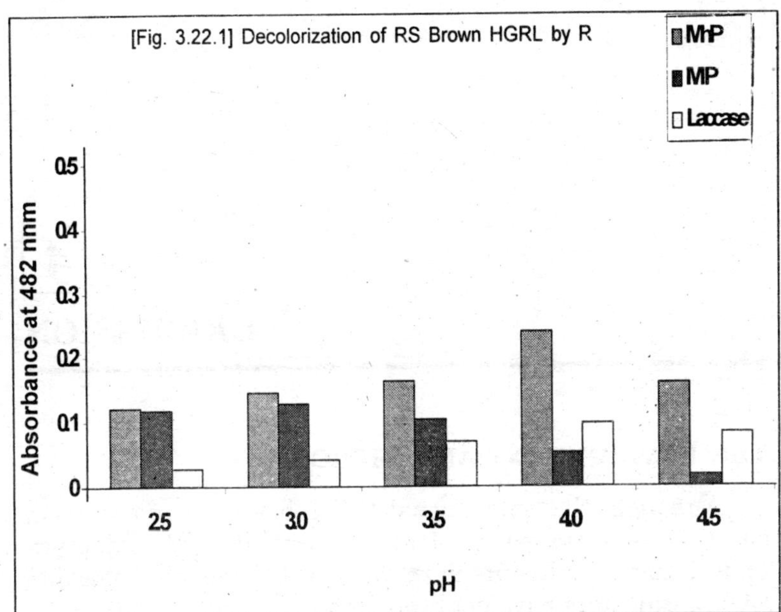
[Fig. 3.22.1] Decolorization of RS Brown HGRL by R
MnP
MP
Laccase
Absorbance at 482 nm
0.5
0.4
0.3
0.2
0.1
0
2.5
3.0
3.5
4.0
4.5
pH

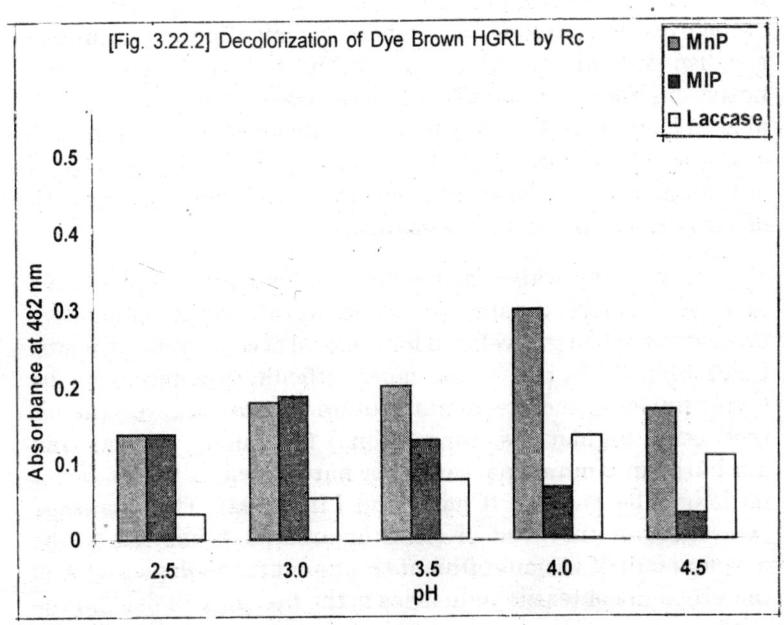
[Fig. 3.22.2] Decolorization of Dye Brown HGRL by Rc
MnP
MIP
Laccase
Absorbance at 482 nm
0.5
0.4
0.3
0.2
0.1
0
2.5
3.0
3.5
4.0
4.5
pH

4

DISCUSSION

THE ENVIRONMENT AND MICROBES

Although, the conventional CETP bas been in operation since 1995, installed to treat the organic load and other undesirable constituents in the wastewater, yet the color and COD parameters have not been found to change significantly.

Reasons for the inadequacy in the treatment might be due to one of the important phase during the treatment process was aerobic (oxidation) conditions where the microbial communities perform critical role. Particularly, the failure to exploit the native microbes adapted to the local conditions might not be able to decolorize dyes due to some unfavorable parameters to support sufficient growth. Thus, the physico-chemical environments may be insufficient to provide adequate growth of various groups of microorganisms.

Color in any water (domestic or industrial) is undesirable and is therefore considered as pollutant. Many innovative processes are being developed for removal of color from effluents. Color removal is one of the most difficult task faced by the textile finishing and dye manufacturers. Due to the government's restrictive regulations, conventional treatment systems and discharge into municipal sewers or natural water bodies would no longer be possible (Chang and Lin, 2002). The increased awareness of the need of clean environment has led to the development of various effluent treatment technologies. One of the problems of textile industries is the presence of dyes in the wastewater. This makes the contamination of environment. Some of the dyes are toxic as well as carcinogenic also. Therefore, there is a need to treat wastewaters for removal of dyes (color) before their discharge in the environment. The dyes are stable to light and oxidizing agents and sometimes are even resistant to bio-oxidation. Textile dyes are of environmental interest because of their potential for formation of toxic aromatic amines during degradation.

In response to this environmental threat, several research laboratories are now attempting to explore and establish systems to biremediate textile effluents. Also that economic removal of dyes from the effluent is becoming increasingly necessary for the industries to mitigate new regulations. Several technologies are available for removal of dyes from the wastewater. There are different treatment options available for removal of dyes from the textile effluents (Jjandra and Loosdrichi, 1997):

1. Ozone treatment—expensive and highly technical approach
2. Low pressure ultra filtration—expensive and highly technical approach
3. Adsorption on activated carbon or alumina
4. Adsorption on waste natural material like peat, banana pith
5. Coagulation, precipitation
6. Biological processes, single, aerobic and anaerobic

The conventional aerobic biological processes can remove 85-90% of BOD/COD from effluents but color removal is only 10-30%. Biological color removal may not be always successful because of the other toxic chemicals existing in the effluents. Also, there are many categories of dyes and biodegradation may not be possible for all. Biological techniques can be cheaper and have more relevance than high technologies in developing and under developed countries.

Reactive dyes are particularly difficult to degrade by conventional methods as they are not readily adsorbed onto the activated sludge biomass (Zaoyan *et al.*, 1992) where they could be degraded.

Azo dyes were considered to be xenobiotics compounds for which biodegradation is difficult but not impossible. For instance, the white rot fungus, Phanerochaete chrysosporium, can mineralize all sulfonated azo dyes. Thus, lignocellulolytic fungi and bacteria can be used for biodegradation of azo dyes.

Anaerobic system to decolorize different azo dyes has been studied in detail (Tjandra and Loosrichi, 1997). They studied three reactive azo dyes, Remazol Red RB, Reactive Red and Remazol Black B and reported decolorization of the first two dyes. Tokyo Institute of Technology group has found new bacteria that do the job economically by decomposing dyestuffs into colorless compounds. The bacteria that they found are easy to culture and thrive in pH range of 4-7.

Thus, biological treatment systems may have promising applications for the removal of color of the azo dye compounds since it is widely reported that azo dyes are gratuitously reduced by anaerobic sludge, anaerobic sediments and anaerobic bacterial enrichments cultures (Brown and Hamburger, 1987). On the other hand, azo dyes are resistant to oxygenolytic attack. Degradation of more than 100 dyes in aerobic activated-sludge systems has been examined by Pagga and Brown (1986), and Shaul *et al.* (1991) found that only a few of them were actually biodegradable.

In the present study, the sampling site selected was CETP of Jetpur Textile Industries's Association. The wastewater received in CETP contained mixture of several chemicals, from inorganic to organic compounds, some of the organic compounds, are easily biodegradable. However, the effluent also possessed mixture of dyes (mostly reactive azo dyes).

Analysis of the samples collected from different stages of CETP has indicated that the composition of the effluent is highly complex. The dyes used in the textile processing units are highly water-soluble and thus separation and purification of individual dye from the samples is impossible. Only the estimation indirect parameters such as color, COD and BOD would suggest the possibility of removal of dyes and color from the effluent.

In view of the above mentioned background and the existing unsolved or partially solved problems of decolorization and degradation of dyes have been considered in this work. Microorganisms are excellent agents that utilize varieties of inorganic and organic molecules (natural or synthetic) for their requirements of nutrients and energy. It is also worth to mention that microbes have already coexisted themselves and with other living organisms since billions of years in the dynamic environment containing immense variety of organic compounds either synthesized by them (including higher organisms, biologically originated) or formed via diagenetic processes. Thus, microbes have been interacting with such vast diversity of potential substrates occurring in their habitats have led to adapt themselves by evolving biocatalyst capable to assimilate, transform, and catabolize such molecules (Jensen, 1976). Since 1950s, a good number of microorganisms have been isolated from natural habitates or man-made habitates (contaminated by human activities). For instance, discharge of wastes by human actions, either directly to any components of environment or into semi-natural bioreactor, *i.e.*, effluent treatment systems. These microbes have been studied to measure their potentials to degrade several xenobiotic compounds that were never encountered with them in their habitates in the recent past. Several workers have examined and reported that microbes have great potential to decolorize and degrade all the dyes present in the effluents.

Recent advances in molecular biology have further strengthened the understanding on the subject.

BACTERIAL ISOLATES

Isolation

All bacteria require an exogenous energy source for their growth. The individual species varies tremendously in the array of compounds utilized for this purpose. Hetetrotrophic bacteria derive energy from the oxidation of or dissimilation of reduced carbon compounds. Amongst the major elements required are carbon, nitrogen, phosphorous and sulfur for cell synthesis. Organic compounds generally serve a dual purpose for bacteria; as a source of energy and as a supply of carbon and other elements. Prokaryotes possess a great metabolic diversity that can use a wide range of carbon sources, from molecules as simple as CO_2 to those complex or unusual as high molecular weight hydrocarbons and recalcitrant-xenobiotics pesticides.

For isolation of bacteria in laboratory several simple to complex molecules of biological origin are employed. Along with such nutritional ingredients, some selective agents are also used for selective enrichment of desired organisms.

The use of peptone or similar hydrolysates in the media also satisfies the nitrogen requirements for organisms incapable of assimilating inorganic nutrients.

Most of bacteria can synthesize essential sulfur compounds from sulfate provided as various inorganic salts. Several studies are available on the inorganic ions requirements of bacteria.

With complex media, many of the ions are generally present as contaminants in medium components such as peptones and yeast extracts and no supplementation is required for routine growth (Hughes and Poole, 1991).

The enrichment medium has been designed and employed was to support growth of existing hetetrotrophic populations. The screening (selective) factor(s) added in the medium was a solution of mixture of 20 dyes (Table 2.2) of known concentration of each dyes, assuming that only those would be favored which were not inhibited by the dyes. Also that the medium consisted of highly nutritious carbon, nitrogen, phosphorous and other required growth factor, *i.e.*, glucose + yeast extract. All the isolation experiments were carried out in closed systems-batch cultures in flask under stationary and shaking conditions. During the isolation and screening processes, it was noticed that decolorization was significantly favored under static culture flasks in which medium supported development of mixed populations of several groups of bacteria. (Table 3.13).

A newly isolated, hetetrotrophic bacterium should be cultivated on a complex medium that supports good, sustained growth, and retains a cellular morphology reflective of the organism in its natural habitates.

In the present study, enrichment, isolation technique employing various mineral salts, glucose and yeast extract as carbon sources, and neutral pH of the medium. The medium did contain 0.02% of mixture of dyes, was used not as selective agent, but to isolate organisms having ability to decolorize dyes. Also to be noted that throughout the study, the isolates obtained were maintained on the same solid medium (Complete Medium Broth + dye mixture).

During the enrichment process, a mixed culture having capacity to decolorize the dye mixture observed were considered as predominant microbial isolates and therefore further attempted to obtain them in pure cultures. Thus the individual bacterial isolates were further screened in the same medium with the dye mixture. Since there might be great variation, in terms of degree and time, to decolorize dyes, the individual isolates have been further assessed that reduced the number of the bacterial isolates from 55 to 30 (Table 3.2 and 3.3).

Successful isolation of a given organism into pure culture requires a sufficiently high proportion of that organism in the sample. If the organism is not numerically dominant, enrichment methods are designed to increase the relative numbers.

Tentative Grouping of the Isolates

In identifying bacteria, certain general characteristics are of primary importance for determining the major group to which the new isolate is most likely to belong. Certain morphological features should be determined:

- Gram's reaction
- Cell morphology
- Presence of special morphological features

Also, three physiological tests are of primary importance:

- The oxidase test
- The Catalase test
- Determination of whether sugar oxidized of fermented

Presence of special physiological characters such as production of pigments may narrow the field of possibilities considerably. Thus, organisms screened have been identified based on phenotypic characters, using an orderly approach. Tables and major keys given in Bergey's Manual of Systematic Bacteriology have been used to compare the results. The study restricted only upto generic level (Table 3.13).

All the thirty bacterial isolates were selectively enriched and employed to confirm for their decolorization potential. Decolorization of color of dye in the medium is the screening experiments. The study carried out did not attempt for detail identification of the isolates, since the purpose of the study was to explore biological agents that remove color of the dye under laboratory conditions and also their efficiency.

COMPARISON OF THE BACTERIAL STRAINS FOR THEIR DYE DECOLORIZATION POTENTIAL

The thirty isolated bacterial strains (Table 3.13) and on the basis of degree of color removed, *i.e.*,% of dye decolorized during fixed time intervals were considered to be more potential. The experiment designed was, thus, enabled to show that only thirteen bacterial strains are reflecting better efficiency to remove color within 24 to 72 hours period.

The tables given below shows list of organisms and% removal of color of individual dye in the CMB medium within 72 hours by selected 13 bacterial isolates:

These thirteen isolates were belonging to distinct categories, both Gram +ve and Gram –ve and their tentative identification indicated that they were the strains belonging to genera *Acetobacter, Pseudomonas, Azotobacter, Bacillus, Legionella, Paracoccus, Enterobacter* and *Xanthomonas*.

Culture No.	Name of the strain	Dye decolorized (%) after 72 hours									
		Jd 0	Jd 1	Jd 2	Jd 3	Jd 4	Jd 5	Jd 6	Jd 7	Jd 8	Jd 10
Kot-01	*Acetobacter sp.*	86	72	85	67	56	72	38	66	66	66
Kot-02	*Pseudomonas sp.*	70	61	85	63	54	69	39	59	56	65
Kot-03	*Azotobacter sp.*	87	74	66	64	47	68	51	64	57	62
Kot-04	*Bacillus sp.*	73	67	76	70	55	72	49	62	68	61
Kot-05	*Bacillus sp.*	98	92	85	64	61	64	48	53	61	68
Kot-06	*Legionella sp.*	84	67	81	69	52	65	52	68	64	59
Kot-07	*Paracoccus sp.*	89	72	81	62	55	62	47	66	65	75
Kot-08	*Enterobacter sp.*	93	78	80	65	53	68	48	67	59	62
Kot-09	*Bacillus sp.*	94	90	80	63	52	63	41	65	58	53
Kot-10	*Xanthomonas sp.*	92	84	70	64	50	62	59	59	63	26
Pseu-I	*Pseudomonas sp.*	95	91	85	65	61	58	48	49	39	55
Pseu-II	*Pseudomonas sp.*	97	96	90	61	66	51	49	52	35	51
Pseu- III	*Pseudomonas sp.*	94	92	40	54	62	56	51	53	71	24

Culture No.	Name of the strain	Dye decolorized (%) after 72 hours									
		Jd 12	Ad 13	Ad 14	Ad 15	Ad 16	Ad 17	Ad 19	Kd 22	Kd 23	Kd 24
Kot-01	*Acetobacter sp.*	49	66	32	47	81	51	65	49	55	50
Kot-02	*Pseudomonas sp.*	47	65	41	38	71	46	69	38	59	54
Kot-03	*Azotobacter sp.*	71	51	25	46	82	49	58	47	61	48
Kot-04	*Bacillus sp.*	57	56	35	39	77	58	62	38	68	70
Kot-05	*Bacillus sp.*	56	64	39	52	58	57	53	32	63	69
Kot-06	*Legionella sp.*	65	52	42	56	82	66	64	31	58	62
Kot-07	*Paracoccus sp.*	45	54	46	55	75	63	67	44	59	68
Kot-08	*Enterobacter sp.*	57	65	54	68	80	58	58	45	49	48
Kot-09	*Bacillus sp.*	59	35	55	71	61	54	62	56	58	59
Kot-10	*Xanthomonas sp.*	65	68	58	70	82	59	59	52	57	37
Pseu-I	*Pseudomonas sp.*	88	78	49	86	82	88	85	86	86	87
Pseu-II	*Pseudomonas sp.*	83	81	37	83	86	80	69	78	84	82
Pseu- III	*Pseudomonas sp.*	79	88	50	87	71	87	77	89	83	87

Further, detailed investigation on the 13 strains when subjected to only three selected dyes showed that the strains Pseudomonas species were capable of decolorizing the three dyes. However, their actions on the three dyes were varying, *i.e.*, RS Brown HGRL was less strongly decolorized than the other two within 24 hours as shown in the following table:

Out of 13 strains, and the (mentioned in above table) the bacterial strain Pseu-II and dye RS Red H5BL selected due to higher and rapid process of decolorization

Culture	Name of strain	Dye decolorization (%) after 24 hours		
		Kemifix Red F6B	RS Red H5BL	RS Brown HGRL
Kot-01	*Acetobacter sp.*	46.86	45.20	34.63
Kot-02	*Pseudomonas sp.*	24.10	23.34	40.83
Kot-03	*Azotobacter sp.*	55.65	54.14	33.25
Kot-04	*Bacillus sp.*	28.92	27.43	40.36
Kot-05	*Bacillus sp.*	90.45	79.98	42.45
Kot-06	*Legionella sp.*	57.72	55.90	36.32
Kot-07	*Paracoccus sp.*	61.12	59.78	44.21
Kot-08	*Enterobacter sp.*	57.34	56.99	31.85
Kot-09	*Bacillus sp.*	81.05	80.88	40.77
Kot-10	*Xanthomonas sp.*	70.64	70.38	44.44
Pseu-I	*Pseudomonas sp.*	90.01	86.35	41.15
Pseu-II	*Pseudomonas sp.*	91.43	89.77	50.77
Pseu-III	*Pseudomonas sp.*	86.45	86.72	52.59

PHYSIOLOGICAL MANIFESTATIONS OF THE ISOLATES AND CULTURAL PARAMETERS WITH RESPECT TO DECOLORIZATION

Physiology is the science of the processes of life. Knowledge on microbial physiology is very extensive and can be manifested in a defined nutrient culture in which it is exponentially growing. Microbes are the basic tools in understanding life processes, since they can be easily grown and manipulated. Also, those different groups of prokaryotes have adapted to grow and multiply in different environments to the extent that they can be isolated from almost any ecological niche, characterized by their very great metabolic diversity. Therefore, it is interesting to understand physiology of microorganisms both in the way in which the growth rate is affected by different parameters, and in the mechanisms by which the cell respond or adapt to stress or extreme/or specialized man-made environment (Dawes and Sutherland, 1992).

Many specific kinds of microorganisms can be obtained from natural or semi-natural habitates (CETP) by creation of an artificial environment for them in the laboratory, which might enhance their growth over competiting organisms. Morphological and/or physiological characteristics of desired organisms, which might give them special advantages over others, are exploited in the formulation of culture media and the choice of incubation medium (www.bact.wisc.edu).

In general, it was aimed in the present study to achieve the following conditions in the laboratory that:

- Satisfy the nutritional requirements of the desired organisms such that they grow well
- Inhibit as many other organisms as possible
- Enhance the detection of desired organisms as early in the process as possible

The source samples inoculated directly into the CMB medium that encouraged the proliferation of desired organisms. For proliferation of organisms grown under medium designed by adding glucose, YE and salts, the batch culture stationary and shaking (100 RPM) conditions permitted growth. The observations during enrichment processes clearly reflected that only the stationary cultural conditions made possible to remove the color of the dye while the shake flasks did not influence decolorization effectively or significantly. Of course, the shake flasks did show growth in the medium. Obviously, removal of color of dyes in the inoculated medium could be possible only after reaching adequate growth of the microbial populations when oxygen was sufficiently present and decolorization initiated as depletion of oxygen occurred in the growth broth. Moreover, the literature indicates the potential of anaerobic systems for non-specific decolorization of water-soluble azo dyes (Brown and Laboureur, 1983).

GENERATION OF ANAEROBIC CONDITION AND THE DECOLORIZATION OF DYES BY THE BACTERIAL STRAINS

Growth of Bacteria in Batch Culture

When a suitable medium is inoculated with a pure culture of bacteria and kept under constant conditions, the organism achieves exponential phase within a short period (6 to 18 hours). There are several factors that govern the behavior of growth pattern of bacteria. The growth rate starts decreasing as various conditions, nutrient and oxygen, changes in pH, and chemical environment of the medium.

In the present study, the process of growth of the bacterial strains, when cultivated in flasks under static and agitating conditions, was carried out with simultaneous observation of decrease in color in the medium, during course of time. As explained earlier, there was no remarkable decolorization recorded when culture flasks were maintained under shaking condition, while in flasks under stationary conditions, a high percentage of decolorization was observed. The major difference is the two physical conditions of cultivation of bacteria was that there was continuous air (oxygen) mass transfer into the medium when flasks were agitated. It follows that batch system could support cell multiplication for only a limited time and with progressive changes in the original medium and environment, biomass increment become slow. Thus, during the stationary growth phase and later on, the cells did favour decolourization of dye(s).

Oxygen Supply in Static Culture Flask Condition

Unlike most nutrients, oxygen is relatively insoluble in water (< 10 mg/l) may quickly become limiting in liquid bacterial cultures. Unless special arrangements are made to ensure that it is supplied with dissolved oxygen continuously during the growth. For the availability of gaseous oxygen to the gas/liquid interface, the opening of the culture vessel must be sufficiently large and stopper or covered with porous material so as to allow maximum exchange of the atmosphere inside and outside the flasks. Under static condition, the liquid culture flask with high liquid volume (100/ml in 250 ml NM Erlenmeyer flask) decreased the gas transfer efficiency and hence the flask achieves microaerophillic to facultative conditions within liquid culture medium (8-12 hrs). The oxygen solution ratio in the flask under static condition decreases rapidly, as the volume of culture medium is high and as the cell mass increases. Also, the partial pressure of oxygen gas in the ambient environment of flask decreases that limits the dissolution of the gas in the solution (Breznak and Coshlow, 2000).

O_2 is a insoluble gas in water at 20°C, in contact with air dissolving only about 9 mg/l. As the temperature is raised oxygen, like any other gas, becomes less soluble. The concentration of dissolved oxygen is also affected by the presence of dissolved salts decreasing with increasing dissolved salt concentration. According to Henry's law, the solubility of oxygen is substantially independent of the total gas pressure but directly proportional to the gaseous phase. The low solubility of oxygen is therefore, of profound importance in considering the supply of oxygen to respiring cultures.

The rate of dissolution of oxygen in a cultural solution is profoundly affected by the characteristic of the mash. Further, these characteristics, change during the course of the growth. Changes, which are likely to occur, are in such properties as viscosity and surface tension. Both of these affect aeration efficiency. The physical chemical properties of cultural solution change continually throughout growth as the concentration of cell mass increases.

The use of shake cultures instead of static cultures has obvious disadvantages when dealing with facultative anaerobic bacteria in this type of sludge since shaking would increase the rate of oxygen absorption in to the culture fluid. Experiments conducted under stationary and shake flash cultivation clearly indicated that the rate of decolorization were significantly higher in stationary cultures.

Several groups of microorganisms reported to decolorize azo dyes are facultative anaerobic bacteria (Chung and Stevens, 1993). Thus, in batch reactor, only the stationary batch system favored decolorization when the gaseous phase of the medium was lacking necessary partial pressure of oxygen. The soluble oxygen in the medium also became depleted and the system had achieved maximum growth and entered into stationary phase.

Similar studies by other researchers showed decolorization of azo dyes under anaerobic conditions (Allan and Roxon, 1974; Brown, 1981, Chung *et al.*, 1978; Beydilli *et al.*, 1998; Brown and Laboureur, 1983; Carliell *et al.*, 1994 and Donlon *et al.*, 1997).

The facultative anaerobes are those organisms that can grow either in the presence or in the absence of air, provided proper nutrients are available in the medium. The category includes organisms that have an anaerobic type of metabolism but are not sensitive to oxygen, as well as organisms that, like *E. coli* or yeast, can shift from aerobic to anaerobic metabolism depending on environmental conditions. A great many bacteria are facultative anaerobes. Such organisms may carry out respiration and fermentation simultaneously, or may shift their metabolism towards one or the other reaction, depending on the availability of electron acceptors. The experiments designed in batch system is self-generating aerobic to anaerobic condition with the flasks under static culture. The first condition of biodecolorization of dye in the medium was that the biomass (at 1.0 OD) and generation of anaerobic condition, *i.e.*, during logarithmic phase carbon sources were used up to produce energy and cell mass, consuming the available dissolved oxygen to oxidize organic carbon. The process of decolorization has been observed to be initiated at the bottom of the culture flask and proceeding in upward direction. However, at the liquid-gaseous interface in the flasks the color retained and removed after extended (48 to 72 hours) hours of incubation.

Thus, faster decolorization rates were observed only under anaerobic conditions. In the upper layer of the medium in the culture flasks the available primary carbon source being oxidized by consuming available oxygen and consequently, the penetration of oxygen into the deeper layer of the flasks would reduce. This process resulted into creation of *anaerobic microniches* into the below layers of the medium by facultative bacteria. During the exponential growth of the cells, CO_2 that generated by aerobic respiration would reduce the partial pressure of O_2 in the bulk gaseous phase of the flask and thus, reduced oxygen level in the phase that subsequently influence the medium liquid phase in the flask.

Effects of Nutritional Parameters on Decolorization in the Cultural Medium

The physical-chemical conditions in the culture flask bioreactor were considered to estimate their influence the rates of dye decolorization. When the conditions in the culture media provided as: CMB medium + dye solution + adequate inoculum and under static batch culture, there was a sufficient cell mass formed within 10 to 12 hours. However, decolorization was not initiated until the required level of DO had reduced (note; during the experiment, DO of the culture broths were determined by Winkler's method). Shake culture flasks showed relatively very low rates of decolorization when compared with the stationary culture flask condition.

The results presented in the Figures 3.4, 3.5, and 3.6 showed that facultative condition was the most favorable to initiate and decolorize dye color in the medium. Similar results were reported by Chang and Lin, 2000; Donlon *et al.*, 1997; Tan and Field,

2000; Carliell *et al.*, 1995; Razo-Flores *et al.*, 1997. The results obtained showed clear indication that the organism first grew to their optimum level when favorable aerobic conditions existed in culture flasks. As soon as depletion of oxygen started, the decolorization process was observed that first initiated from the bottom of the medium to upward direction in culture flasks. Supplying oxygen into the shake flask did not support high percentage decolorization.

Tan and Field, 2000 has explained integrated anaerobic/ aerobic conditions in a single bioreactor for the complete mineralization of azo dyes. The observations made in the present study, the static culture flasks acting as bioreactor in which anaerobic-aerobic conditions (bottom to top) generated had made the decolorization of dye possible. The first step in the biodegradation of azo dyes was the anaerobic azo dye reduction. However, faster dye reduction was observed in the bottom of the culture flasks, having maximum anaerobic conditions. Under anaerobic conditions co-substrate acted as electron donor for the azo dye reduction. The anaerobic conditions were unfavorable to further degradation and mineralization.

The experimental results mentioned showed that different nutritional ingredients, such as glucose, yeast extract, peptone, benzoic acid when added in the medium individually and in combinations.

The results clearly indicated that either yeast extract or peptone were required in the medium. Glucose or benzoic acid alone in medium served as carbon and energy source but not providing the factor(s) that enabled decolorization.

From the results it can concluded as:

1. Either yeast extract or peptone when added, almost complete decolorization of the dye was occurred.
2. When glucose or benzoic acid added, there was no observable change of dye color in the medium.
3. It was evident that glucose or benzoic acid served as only carbon/energy source. However, yeast extract or peptone did provide some factor(s) alongwith carbon,

nitrogen, energy that lead to complete decolorization. Work of Chang and Lin (2000) reported that addition of yeast extract significantly enhanced the efficiency of dye decolorization under anaerobic conditions.

As stated by Chang and Lin (2000), yeast extract and/or peptone did serve as the sole supplemental respiration substrate to maintain decolorization activity.

The appropriate conditions in the culture flask bioreactor had thus given rise to enhancement of the dye decolorization, the action carried out by the inoculated organism Pseu-II. Results on biomass data of the inoculated bacterial strain Pseu-II studied, indicating that only carbon sources such as glucose or benzoic acid did not favour good growth rates. Yeast extract or peptone had some influence on the growth showing increase in biomass, while the combinations of yeast extract + peptone or Benzoic acid + yeast extract had significant increase in biomass (6.0 mg/ml and 4.0 mg/ml (Fig. 3.7). The combinations containing either yeast extract or peptone with other carbon source could obtain higher biomass and decolorization also.

In the process of decolorization under anaerobic conditions, the strains inoculated in the flasks containing medium, showed that each strain preferentially acted on dyes. All the 20 dyes employed in the present study in the CMB medium inoculated with each of 30 bacterial isolates separately had shown different affinities for each dye to decolorize ranging from 70% for Jd-00, the highest to 48.00% for Jd-14. The secondary screening assay investigation of 13 bacterial cultures and using only three dyes under static and agitating conditions suggested that the dye RS Red H5BL was decolorized maximally (60 to 95%) by all the strains inoculated in the medium. The other two dyes did not observed to be decolorized so efficiently by these strains (40% to 70%), except the strain kot-5 (88%).

On the basis of the results of the 3 dyes used in this experiment, dye RS Red H5BL showed as a better substrate affinity that decolorized highest percentage, whereas the dye Kemifix Red F6B decolorized slightly lesser and dye RS Brown HGRL decolorized the lowest. Thus, the organisms showed substrate specificity (affinity) that varies with change in the substrate (dye). The assay also indicated that the bacterial strain Pseu-I, Pseu-II, and Pseu-III were showing the highest percentage of decolorization. Thus, the strains of *Pseudomonas* were thought to have highest potential and therefore Pseu-II strain had been used in the further study.

In terms of biodegradation/biodecolorization studies, the main agents are the microbial population occurring in that specific environment. These agents are required to be properly defined with respect to their decolorization capability. Therefore, the small-scale laboratory test systems used in which enrichment culture techniques and screening were performed.

The results of physico-chemical analysis of the textile effluents confirmed that reduction in COD and BOD of the organic load, indicating development of microbial community during the treatment processes in the CETP (Figure 3.1). These results in principle, showed the presence of microorganisms. Before being able to isolate these target organisms, from the samples, it was decided to enrich the organisms. The method followed the steps:

1. To enrich microbial populations containing the potential dye decolorization on rich complex medium that encouraged high microbial cell densities and enabled to cultivate colonies.
2. Then, morphologically different colonies were subcultured on dye containing CMB agar plates and subsequently transferred on the agar slants to maintain.
3. Screening tests were conducted to confirm the isolates that could decolorize dyes in the CMB medium under shaking and static conditions. The tests had confirmed the capabilities of the isolates, through the degree of decolorization potential varied among the bacterial culture isolated.
4. The bacterial isolates were further studied to determine their taxonomic positions based on certain phenotypic characteristics (Table 3.4).

5. Among the 55 isolates obtained during enrichment experiments, 30 isolates were selected, grouped them into their generic positions, and were subsequently screened by measuring degradation potential at 24, 48, and 96 hours intervals of incubations. As stated earlier, the facultative types were observed to be the most potential decolorizers.
6. The stability of the decolorizers was observed throughout the work. It seemed that yeast extract in the medium did favour of decolorization process and therefore yeast extract had been considered as the important nutritional parameter that maintained the stability of the isolates.

For the strain Pseu-II a more detailed decolorization medium optimization was performed (Fig. 3.7). The addition of complex ingredients such as yeast extract and/or peptone was responsible for higher rates of decolorization by the strain.

Most interestingly finding is that glucose alone supported growth of the organism (Pseu-II) but did not favour recognizable decolorization of dye (Fig. 3.7). The glucose molecules were able to provide carbon and energy source but not nitrogen source. In presence of yeast extract or peptone, the organism showed equal potential to decolorize dye, reflecting that either growth factor(s) or nitrogenous compound(s) were favoring decolorization (Fig. 3.4). Generalizing this, it can be stated that the decolorization process is dependent on either yeast extract or peptone. The process does not favour by the only readily available carbon sources such as glucose or benzoic acid.

Comparison of the decolorization results obtained with 30 isolates, 13 isolates show initiation of decolorization process after 14-18 hours of incubation. In all the cases more than 50% decolorization occurred within 24 hours. The interesting point be noted is that depletion oxygen partial pressure in the medium favored initiation of the process in all the cases. This indicates that facultative condition improved the process.

ALTERNATIVE EVIDENCE SHOWING DECOLORIZATION OF DYE(S)

The evidence presented in the plates (01 to 20) showing comparison of decolorization process in presence and absence of the organism with the 20 dyes. These photographs alongwith the TLC results confirmed that the native dye(s) added in the medium disappeared after 72 hours incubation.

Further, the UV-Visible spectroscopic analysis (Figs. 3.11 to 3.17) performed on the samples also definitely proved the TLC results as shown by the disappearance of peaks at the specific wavelength.

Indirect tests were followed using ring cleavage reaction (*Rothera test*), which showed positive results indicating decolorization of dyes and aromatic ring cleavage of the dye structure, suggesting further catabolism of the dye.

INDUCTION OF RING CLEAVAGE ACTIVITY IN PRESENCE OF DYES CHARACTERIZATION OF RING CLEAVAGE REACTION

Many bacteria and fungi degrade aromatic rings. Initially the rings are hydroxylated by monooxygenases enzymes, forming catechol (aromatic compounds with two hydroxyl groups in adjacent position) and are then cleaved by diooxygenases to yield muconic acids or muconic semialdehyde.

Rothera reaction has been used to indicate ring cleavage of aromatic molecules, yielding the product α-ketoadipate. The bacterial strain Pseu-II was used to determine occurrence of ring cleavage system in the organism, when grown in the presence and absence of the dyes in the medium.

It is well known that aromatic hydrocarbons and their substituted derivatives must be modified into o-diphenols through different peripheral reactions before ring cleavage can take place. Many bacteria, mainly belonging to genus *Pseudomonas* can carry out ring fission that subsequently be reacted for energy-yielding processes (Galli, 1994).

The dyes used might be best effectors (inducers) but inhibitory substrate for the enzyme(s). The results obtained in this experiment, did show the presence of enzymes activity as positive *meta* cleavage by Rothera's reaction.

Question does arise from this results is that whether the dyes were cleaved or simply worked as effectors and induced the necessary enzyme(s). Thus, the results does not confirm that ring cleavage reaction occurred on the dyes. There might be possibility of ring cleavage and hence the degradation.

RING CLAVAGE OF PAHs

The growth of these isolates in the MM consisting of various PAHs indicates the utilization of these PAHs as source of carbon in presence of co-substrate such as glucose. A method originally suggested by Dr. K. Hosokawa (Stanier *et al.*, 1966), by which it can be tested the *ortho* or *meta* cleavage of two central intermediates in the metabolism of aromatic compounds (catechol and protocatechuate).

Usually bacteria utilize extra diol (*meta*) or intra diol (*ortho*) ring fission to cleave aromatic nucleus. Extra (meta) diol ring fission is plasmid coded (Holloway and Morgan, 1986) whereas the latter one is genomic coded one (Harayama *et al.*, 1987). Results show that Pseu-I, Pseu-II show the meta cleavage indicating the presence and action of extra diol ring fission, a plasmid encoded characteristics (Holloway and Morgan, 1986). At same time, Pseu-III, shows the ortho cleavage indicating the presence and action of intra diol ring fission, a chromosomally encoded characteristics (Harayama *et al.*, 1987). Production of the yellow product [α-hydroxy muconic semialdehyde (α-HMS)] from catechol *i.e., meta* ring cleavage and production of β-ketoadipate from catechol *i.e., ortho* cleavage was tested for detecting meta or ortho ring fission pathway .

Out of all PAHs selected for this study, some of them have ubiquitous distribution, *e.g.*, benzo[a]pyrene, a stable, five ring structure (Suess, 1976). Benzo[a]pyrene has been monitored extensively in ambient air and has been detected in ground water, surface water, and vegetation. Benzo[a]pyrene concentrations have also been detected in food, including vegetable oils, meats, fruits, vegetables, grains and beverages such as rum and tea. Although trace amount can be detected in many environmental samples, the highest levels of benzo[a]-pyrene are usually found in soils and sediments. Benzo[a]pyrene is potent carcinogenic and has been extensively and precisely studied.

Other PAHs induce neoplasia and studies were initiated after the identification of benzo[a]pyrene as a major active component of coal tar during the 1930's. Benzo[a]pyrene is also one of the sixteen PAHs on the US EPA,s priority pollutant list and is defined as an environmentally recalcitrant compound, that is, a compound which exhibits structural resistance to biodegradation by the available metabolizing capabilities of soil microbiota.

In the present work, aerobic degradation of various PAHs (Table 2.4) were examined with three strains of Pseudomonas (Pseu-I, Pseu-II, and Pseu-III) showing varying ability to grow in MM medium containing PHAs. The results, when compared with controls, *i.e.*, medium with PAHs but with out glucose (carbon source) and with glucose and, clearly indicate the three strain could cleave the ring structures (either at *ortho* or *meta* position) in presence of primary carbon source, as glucose.

Thus, it is evident that the strains had varying capabilities to attack on different PAHs only when an adequate carbon source was provided in the medium under aerobic condition. Recently many microorganisms have been shown to metabolize the three ring PAHs, phenanthrene and anthracene. The four ring PAHs, pyrene, chrysene and Benz[a]anthracene have also now been shown to be susceptible to microbial biodegradation (Sutherland *et al.*, 1995; Cerniglia, 1992). Less is known about the potential for biodegradation of higher molecular weight PAHs with greater than four rings, such as benzo[a] pyrene, benz[a]anthracene, and fluoranthrene, Acenaphthene etc.

Under certain conditions, benzo[a]pyrene has been shown to be biodegraded by a variety of microorganisms including bacteria, fungi, yeasts, and algae; however, bacteria capable of utilizing benzo[a]pyrene as a sole source of carbon and energy have not yet been demonstrated. However, some microorganisms can oxidize these chemically stable PAHs when grown on an alternative carbon source. The process is known as co-metabolism. As defined, co-metabolism is the gratuitous metabolic transformation of a substance by a microorganism growing on another substrate; the co-metabolized substance is not incorporated into an organism's biomass, and the organism does not derive energy from the transformation of that substance (Atlas and Bartha, 1993)

Based on a review of the recent data, research in a number of areas would be useful for enhancing the bioremediation of PAHs. Microorganisms that possess the ability to degrade higher molecular weight PAHs should be isolated and characterized and genes coding for enzymes involved in higher molecular weight PAHs degradation must be cloned and analyzed. The biodegradation of heterocyclic analogs of PAHs, nitrated PAHs, and alkylated PAHs should be determined. Experimental biodegradation of complex mixtures of PAHs to determine the effect of one PAHs on the biodegradability of another will be of interest. The degradation and metabolic pathways of PAHs in freshwater and marine bacteria should be compared. The mechanisms by which microorganisms degrade bound PAHs substrates and an assessment of the environmental impact of PAHs on coastal ecosystems and the bioavailability and toxicity of the parent compounds and biodegradation products should be relevant to further study.

WHITE ROT FUNGI

White rot fungi *C. polysona* and newly isolated *strain Rc* under the experimental condition employed demonstrated the ability to decolorize various synthetic dyes tested. However, Medium composition has a strong influence on the time and level of expression of ligninolytic enzyme system of WRF. Carbon and Nitrogen—source in cultivation medium determine the composition of ligninolytic system produced by the fungus (Kirk *et al.* 1986).

There are some unique properties known to be associated with ligninolytic peroxidase. The optimum pH for decolorization was between 3.5 to 4.0; it was produced as a part of the ligninolytic system of *C. polysona* and strain Rc during the growth of fungi on wheat straw—a natural lignocellulosics substrate, and is produced contemporarily with other ligninolytic enzymes.

Decolorization of all the dyes observed to be associated with either MnP/MIP or very less by Laccase activity and therefore H_2O_2 dependent. Decolorization of these dyes is oxidative and the color changes observed suggested hypsochromic shift; as the dye was gradually degraded, color of the reaction mixture changed from red to reddish brown to yellow before getting decolorized completely, *i.e.*, λ_{max} shifted to lower wavelengths and with a decrease in molar adsorption coefficient (e). Occasionally, increase in the absorbance (negative% decolorization) of the reaction mixture occurred upon enzymatic degradation of certain dyes suggesting bathochromic shift (reverse hypsochromic shift) perhaps because of the formation of (polymerization products to a chromophore with a more complex conjugation system (Jani, 1997; Patel 1997).

Reflecting the growing reliance on biodegradation processes to clean up contaminated natural resources, numerous strategies and methods are available for meaningful degradation. The choice of a method will depend on the information required as determined by assessment criteria, applicability, and cost. Increased analytical capabilities have resulted in greater extraction efficiencies and diminishing limits of detection.With this expanded capabilities, it is of paramount importance to process and interpret this growing information in its proper environmental context. This means cautious extrapolation to real field conditions with continued development of our understanding of contaminant flux and bioavailability.These two criteria are particularly important because they ultimately will dictate the contaminant's risk benefit dyanimics and play an imortant role in establishing realistic endpoint criteria for cleanup.

REFERENCES

ALEXANDER, M. 1994. Biodegradation and Bioremediation, Academic Press, New York.

ALLAN, R. AND ROXON, J.J. 1974. Metabolism by intestinal bacteria: the effect of bile salts on tartazine azo reduction. Xenobiotica. 4(10):637-643.

ANDREW *et al.*, 1995. (American Public Health Association), APHA 19th edition, Standard Methods for the Examination of water and wastewater. pp. 5-12

ANONYMOUS, 1995. Tentative report: industrial counseling textiles/Environmental Audit, Jetpur/Gujarat/India.

ANSON, J.G. AND MACKINNON, G. 1984. Novel Pseudomonas plasmid involved in aniline degradation. Applied and Environmental Microbiology. 48(4):868-869.

ATLAS, R.M., AND R. BARTHA. 1993. Microbial Ecology: Fundamentals and Applications, 3rd edition. Benjamin Cummings, Redwood City, CA.

AUST, S.D. AND J. T. BENSON. 1993. The fungus among us: Use of white rot fungi to biodegrade environmental pollutants. Environ. Health Perspect. 101:232-233.

AUST, S.D. 1990. Degradation of environmental pollutants by Phanerochaete chrysosporium. Microb. Ecol. 20:197-209.

BAIRD, R., CAMONA, L. AND JENKINS R.L. 1977. Behavior of benzidine and other aromatic amines in aerobic wastewater treatment. Journal WPCF, July:1609-1615.

BASSHAM, J. A. 1975. The substrate: general consideration. In. Wilke, C.R. (ed.) Cellulose as a chemical energy resource. Biotechnology and bioengineering symposium no 5. Wiley, New York, pp. 9-19.

BERNA, J.L., MORENO, A. AND FERRER, J. 1991. The behavior of LAS in the environment. Journal of Chemical Technology and Biotechnology. 50(3):387-398.

BEYDILLI, M.I., PAVLOSTATHIS, S.G. AND TINCHER, W.C. 1998. Decolorization and toxicity screening of selected reactive azo dyes under methanogenic conditions. Water science and technology. 38(4-5):225-232.

BLUMEL, S., CONTZEN, M., LUTZ, M., STOLZ, A. AND KNACKMUSS, H.J. 1998. Isolation of a bacterial strain with the ability to utilize the sulfonated azo compound 4'-carboxy-4' sulfoazobenzene as the sole source of carbon and energy. Applied and Environmental microbiology. 64(6):2315-2317.

BOOMINATHAN, K. R., AND REDDY, C.A. 1992. Fungal degradation of lignin: biotechnological applications. In: Akora, D.K., Elander, R.P., Mukerji, K.G. (eds) Handbook of applied mycology, Vol. 4, Dekker, New York, pp. 763-782.

BOSSERT, I.D., AND BARTHA, R. 1986. Structure biodegradability relationships of polycyclic aromatic hydrocarbons in soil. Bulletin of Environmental Contamination and Toxicology. 37:490-495.

BRENZAK, JOHN, A. AND RALPH N. COSTILOW, 2000. Physicochemical factors in Growth in methods for general and molecular bacteriology. (Philipp Gerhardt, RGE Murray, Willis. A. Wood and Noel. R. Kreig), pp. 137-154

BRETSCHER, R.B. 1981. Waste disposal in chemical industry. In: Microbial degradation of xenobiotics and recalcitrant compounds , Leisinger T., Cook A.M., Hutter, R. and Nuesch, J., Academic Press Inc., London. pp. 65-74.

BROWN, D. AND HAMBURGER, B. 1987. The degradation of dyestuffs part III- Investigation of their ultimate degradability. Chemosphere 16(7):1539-1553.

Brown, D. and Laboureur, P. 1983. The degradation of dyestuffs. Part-I: Primary biodegradation under anaerobic conditions. Chemosphere. 12:397-404.

Brown, D. and Laboureur, P. 1983a. The aerobic biodegradability of primary aromatic amines. Chemosphere. 12(3):405-414.

Brown, D. and Laboureur, P. 1983b. The degradation of dyestuffs: Part 1-Primary biodegradation under anaerobic conditions. Chemosphere. 12(3):397-404.

Brown, J.P. 1981. Reduction of Polymeric azo and nitro dyes by intestinal bacteria. Applied and Environmental microbiology. 41:1283-1286. Naerobes. Applied and Environmental Microbiology. 42:641-6.

Bumpus, J.A., Tien, M., Wright, D. and Aust, S.D. 1985. Oxidation of persistent environmental pollutant by a white rot fungus. Science. 228:1434-1436.

Buswell, J.A. and Odier, E. 1987. Lignin biodegradation. Crit. Rev. Biotechnol., 6:1-60.

Buswell, J.A., Mollet, B. and Odier, E. 1984. Ligninolytic enzyme production by Phanerochaete chrysosporium under condition of nutrient sufficiency. FEMS Microbial Letter. 25: 295-299.

Cain, R.B. and Farr, D.R. 1968. Metabolism of aryl sulphonates by microorganisms. Biochemical Journal. 106:859-877

Cao, W., Mahadevan, B., Crawford, D.L. and Crawford, R.L. 1993. Characterization of an extracellular azo dye oxidizing peroxidase from Flavobacterium sp. NVCC: 39723. Enzyme and microbial technology. 15:810-817.

Carliell, C.M., Barclay, S.J. Naidoo, N., Buckley, C.A. Mulholland, D.A. and Senior, E. 1995. Microbial decolorization of a reactive dye under anaerobic condition. Water SA. 21(1): 61-69.

CARLIELL, C.M., BARCLAY, S.J., SHAW, C., WHEATLEY, A.D. AND BUCKLEY, C.A. 1998. The effect of salts used in textile dyeing on microbial decolorization of a reactive azo dye. Environmental Technology. 19(11):1133-1137.

CARLIELL, C.M., GODEFROY, S.J., NAIDOO, N., BUCKLEY, C.A., SENIOR, E., MULHOLLAND, D. AND MURTINEIGH, B. S. 1994. Anaerobic decolorization of azo dyes, seventh interaction syposium on anaerobic digestion: RSA Litho (Pty) Ltd., Goodwood, South Africa, Capetown, South Africa. pp. 303-306.

CARMICHAEL, L.M. AND PFAENDER, F.K. 1997. Polynuclear aromatic hydrocarbon metabolism in soils: Relationship to soil characteristics and preexposure. Environ. Toxicol.Chem. 16:666-675.

CARMICHAEL, L.M., CHRISTMAN, R.F. AND PFAENDER, F.K. 1997. Desorption and mineralization kinetics of phenanthrene and chrysene in contaminated soils. Environ. Sci. Technol. 31:126-132.

CATALLO, W.J. AND PORTIER, R.J. 1992. Use of indiginious and adapted microbial assemblages in the removal of organic chemicals from soils and sediments. Water Sci. Technol. 25:229-237

CERNIGLIA, C.E. 1992. Biodegradation of polycyclic aromatic hydrocarbons. Biodegradation. 3:351-368.

CERNIGLIA, C. E. AND HEITKAMP, M.A. 1990. Polycyclic aromatic hydrocarbon degradation by Mycobacterium. Methods Enzymol. 188:148-153.

CHANG, JO-SHU AND LIN YU-CHIH, 2000. Fed-batch bioreactor strategies for microbial decolorization of azo dyes using a Pseudomonas luteola strain. Biotechnol. Prog. 16:979-985.

CHINWEKITVANICH, S., TUNTOOLVEST, M. AND PANSWAD, T. 2000. Anaerobic decolorization of reactive dyebath effluents by a two-stage UASB system with tapioca as a co-substrate. Water Research. 34 (8):2223-2232.

CHRISTON J. HURST, RONALD L. CRAWFORD, GUY R. KNUDSEN, MICHAEL J. MCINERNEY, AND LINDA, D. STETZENBACH, 2002. Manual of Environmental Microbiology, 2nd Edition, ASM Press, Washington D.C.

CHUNG, K.T. AND STEVENS, S.E. JR. 1993. Degradation of azodyes by environmental microorganisms and helminth. Environ. Toxicol. Chem. 12(11):2121-2132.

CHUNG, K.T. AND CERNIGLIA , C.E. 1992. Mutagenicity of azo dyes: Structure activity relashionship. Mutation Research. 277:201-220.

CHUNG, K.T., FULK, G.E. AND ANDREWS, A.E. 1981. Mutagenicity testing of some commonly used dyes. Applied and Environmental Microbiology. 42(4):641-648.

COLLINS, P.J. AND DOBSON, A.D.W. 1995. Extracellular lignin and manganese production by the white rot fungi *coriolus versicolor*. 290 Biotechnology Letters (17)9:989-992.

COOKSON, J.T. 1994. Bioremediation Engineering: Design and Application, McGraw-Hill, New York.

COUGHLIN, M.F., KINKLE, B.K., TEPPER, A. AND BISHOP, P.L. 1997. Characterization of aerobic azo dye-degrading bacteria and their activity in biofilm. Water Science and Technology 36(1):215-220.

CRIPPS, C., BUMPUS, J.A. AND AUST, S.D. 1990. Biodegradation of azo and heterocyclic dyes by Phanerochaete chrysosporium. Appl. Environ. Microbiol. 56:1114-1118.

DAWES, I.W. AND SUTHERLAND, I.W., 1992. Microbial Physiology 2nd ed. Blachwell Science.

DE JONG, E., J.A. FIELD AND J.A.M. DE BONT 1994. Aryl alcohols in the physiology of ligninolytic fungi. FEMS Microbiol. Rev. 13:153-188.

DIPPLE, A., CHENG, S.C. AND BIGGER, C.A.H. 1990. Polycyclic Aromatic Hydrocarbons Carcinogens. In: Mutagens and Carcinogens in the Diet. Edited by Pariza M.W., Aeschbacher HU, Felton J.S. and Sato S. New York: Wiley-Liss. 72:141-157.

DONLON, B.A., RAZO-FLORES, E., LUIJTEN, M., SWARTS, H., LETTINGA, G. AND FIELD, J.A. 1997. Detoxification and partial mineralization of the azo mordant orange 1 in a continuous up flow anaerobic sludge-blanket reactor. Applied Microbiology and Biotechnology. 47(1):83-90.

EASTON, J.R. 1995. The dye maker's view. In: Colour in dye house effluent. Cooper P., Society of Dyers and Colorists, Nottingham. pp. 9-22.

EATON, R.A. AND HALE, M.D.C. 1993. Wood, decay, pests and prevention. Chapman and Hall, London.

EDAKA, E., OGAWA, T. AND HORISTSU, H. 1982. Degradative pathway of p-aminoazobenzene by Bacillus subtilis. European Journal of Applied Microbiology. 15:141-143.

EDAKA, F., OGAWA, T., SAKAGUCHI, M. AND TAKIHI, N. 1980. Characteristics of Bacillus subtilis azoreductase. Research Report of faculty Engineering, Gifii University. 30: 53-58.

EFROYMSON, R. AND ALEXANDER, M. 1991. Biodegradation of Phenanthrene by Arthrobacter species of hydrocarbons partitioned in to an organic solvent. Appl Environ Microbiol. 57:1441-1447.

EGGERT C. TEMP, U. AND ERIKSSON, K.E. 1996. The ligninolytic system of the white-rot fungus Pycnoporus cinnabarinus; purification and characterization of the laccase. Appl Environ Microbiol. 62:1151-1158.

FARRELL, R.L., K.E. MURTAGH, M. TIEN, M.D. MOZUCH, AND KIRK, T.K. 1989. Physical and enzymatic properties of lignin peroxidase isoenzymes from Phanerochaete chrysosporium. Enzyme Microb.Technol.11:322-328.

FEDERLE, T.W. AND VENTULLO, R.M. 1990. Mineralization of surfactants by the microbiota of submerged plant detritus. Applied and Environmental Microbiology. 56(2):333-339.

FIELD, J.A., STAMS, A.J.M., KATO, M. AND SCHRAA, G. 1995. Enhanced biodegradation of aromatic pollution in co-culture of anaerobic and aerobic bacterial consortia. Antonie van Leeuwenhoek. 67:47-77.

FIELD, J.A., E.DE JONG, G. FEIJOO COSTA, AND J.A. M.DE BONT 1993. Screening for ligninolytic fungi applicable to the biodegradation of xenobiotic. TiBTech.11:44-49.

FOCHT, D. D. AND WILLIAMS, F. D. 1970. The degradation of p-toluenesulfonate by a Pseudomonas. Canadian Journal Microbiology. 16:309-316.

FUTOMA, D. J., S. R. SMITH, T. E. SMITH AND TANAKA, J. 1981. Solubility studies of PAH in water, Pp: 13-24. In: D. J. Futoma, S. R. Smith, T. E. Smith and J. Tanaka (ed), Polycyclic aromatic hydrocarbons in water systems. CRC press, Boca Raton, Fla.

GALLI, E. 1994. The role of microorganisms in environmental decontamination. Pp.235-246. In: A. Renzoni, N. Mattei, L. Lari, and M.C. Fossi (ed.), Contaminants in the Environment. CRC Press, Inc.,Boca Raton, Fla.

GHISALBA, O.P. CEVEY, M. KUENZI AND SCHAR, H.P. 1985. Biodegradation of chemical waste by specialized methylotrops, and alternatives to physical methods of waste disposal. Conserv. Recycling. 8:47-71

GIBSON, D.T. 1991. Biodegradation, biotransformation and the Belmont. J. Ind. Microbiol. 12:1-12.

GIBSON, D.T. 1984. Microbial degradation of organic compounds. New York: Marcel decker.

GIBSON, D.T. AND SUBRAMANIAN, V. 1984. Microbial degradation of aromatic hydrocarbons. pp: 181-252. In: D.T. Gibson (ed.), Microbial biodegradation of organic compounds. Marcel Dekker, New York.

GINGELL, R. AND WALKER, R. 1971. Mechanisms of azo reduction by streptococcus faecalis II. The role of soluble flavins. Xenobiotica. 1(3): 231-239.

GLENN, J.A., MORGAN, M. A., MAYFIELD, M.B., KUWAHARA, M. AND GOLD, M.H. 1983. Biochemical and biophysical Research Communications. 11:1077-1083.

GLENN, J.K., L. AKILESWARAN AND GOLD, M.H. 1986. Mn(II) oxidation is the principle function of the extracellular

Mn-peroxidase from Phanerochaete chrysosporium. Arch. Biochem. Biophys. 251:88-696.

GOSZEZYNSKI, S., A. PASZCZYNSKI, M.B. PASTY-GRYGSBY, R.L. CRAWFORD AND CRAWFORD, D.L. 1994. New pathway for degradation of sulfonated azo dyes by microbial peroxidase of Phanerochaete chrysosporium and Streptomyces chromofuscus. J. Bacteriol. 176:1339-1347.

GOVINDSWAMI, M., SCHMID T.M., WHITE, D.C. AND LOPER, J.C. 1993. Phylogenetic analysis of bacterial aerobic degradation of azo dyes. Journal of Bacteriology. 175:6062-6066.

GREGORY, P. 1993. Dyes and dye intermediates. In: Kroschwitz, J.I.(ed) Encyclopedia of chemical technology, vol. 8. Wiley, New York. pp. 544-545.

GRIFOLL, M., CASELLAS, M., BAYONA, J.M. AND SOLANAS, A.M. 1992. Isolation and characterization of a fluorene-degrading bacterium: Identification of ring oxidation and ring fission products. Appl. Environ. Microbiol. 58:2910-2917.

GROSSER, R.J., WARSHAWSKY, D. AND VESTAL, J.R. 1991. Indigenous and enhanced mineralization of Pyrene, Benzo (a) pyrene and carbazole in soils. Appl Wnviron Microbial. 57:3462-3469.

GUTHRIE, E.A. AND PFAENDER, F.K. 1998. Reduced pyrene bioavailability in microbially active soils Environ. Sci. Technol. 32:501-508.

HALL, M. AND GROVER, P.L. 1990. Polycyclic aromatic hydrocarbons: metabolism, activation, and tumor initiation. pp. 327-372. In: C.S. Cooper and P.L. Grover (ed.), Chemical Carcinogenesis and Mutagenesis I. Handbook of Experimental Pharmacology, Vol.94/I. Springer Verlag, NewYork.

HAMMEL, K.E. 1992. Oxidation of aromatic pollutants by lignin degrading fungi and their extracellular peroxidases. In: Siegel H. and Siegel A. (ed.) Metal ions in Biological Systems. Vol. 28, Degradation of environmental pollutants by MOS and their metalloenzymes. Marcel Dekker, New York. pp. 41-60.

HANSEN, C., FORTNAGEL, P. AND WITTICH, R.M. 1992. Initial reactions in the mineralization of 2-sulfobenzoate by Pseudomonas sp. RW611. FEMS Microbiology letters. 92(1): 35-40.

HARVEY, P.J., H.E. SHOEMAKER AND PALMER, J. M. 1986b. Veratryl alcohol as a mediator and the role of radical cations in lignin biodegradation by Phanerochaete chrysosporium. FEMS Microbiol. Lett. 13:125-135.

HARVEY, P.J., SCHOEMAKER, H.E. AND PALMER, J.M. 1985a. Annual Proceedings of the Phytochemical Society of Europe. 26:249-266.

HARYAMA, S., REKIK, M., WASSERFALLEN, A. AND BAIROCH, A. 1987. Evolutionary relationships between catabolic pathways for aromatics: Conversation of gene order and nucleotide sequences of catechol oxidation genes of pWWO and NAH7 plasmids. Mol. Gen. Genet. 210:340-343.

HATAKKA, A. 1994. Lignin modifying enzymes from selected white rot fungi: Production and role in lignin degradation. FEMS Microbiol. Lett. 13:125-135.

HATAKKA, A. AND UUSI-RAUVA, A. K. 1983. Degradation of ^{14}C-labelled poplar wood lignin by selected white rot fungi. Eur. J. Appl. Microbiol. Biotechnol. 17:235-242.

HAUG, W., SCHMIDT, A., NORTEMANN, B., HEMPEL, D.C., SLOTZ, A. AND KNACKMUSS, H.J. 1991. Mineralization of the sulfonated azo dye Mordant Yellow 3 by a 6-aminonapthalene-2-sulfonate-degrading bacterial consortium. Applied and Environmental Microbiology. 57(11):3144-3149.

HEINFILING, A, MARTINEZ, M. J, MARTINEZ, A. T. AND BERGBAUER, M. S. U. 1998. Transformation of industrial dyes by manganese peroxidases from *Bjerkandera adusta* and *Pleurotus eryngii* in a manganese independent reaction. Appl Environ Microbiol. 64:2788-2793

HEISS, G.S., GOWAN, B. AND DABBAS, E.R. 1992. Cloning of DNA from a *Rhodococcus* strain conferring the ability to decolorize sulfonated azo dyes. FEMS Microbiology Letters, 99(2-3):221-226.

HEITKAMP, M.A., FREEMAN, J.P., MILLER, D.W. AND CERNIGLIA, C.E. 1988. Pyrene degradation by *Mycobacterium* sp. Identification of ring oxidation and ring fission products. Appl. Environ. Microbial. 54:2556-2565.

HEITKAMP, M.A. AND CERNIGLIA, C.E. 1987. The effects of chemical structure and exposure on the microbial degradation of polycyclic aromatic hydrocarbons in freshwater and estuarine ecosystems. Environmental Toxicology and Chemistry. 6:535-546.

HIGUCHI, T. 1986. Catabolic pathways and role of ligninases for the degradation of lignin substructure models by white rot fungi. Wood res. 73:58-81.

HIGUCHI, T. 1987. Biochemistry of lignin and its potential application. Proc. 4th Intern. Symp. Wood Pulping Chemistry, Vol. 1: 27-30 April, Paris. pp. 133-138.

HOLLOWAY, B.W. AND MORGAN, A.F. 1986. Genome organization in Pseudomonas. Annu. Rev. Microbiol. 40,79.

HOOPER, S.W. 1991. Biodegradation of sulfonated aromatics. In: Biological degradation and bioremediation of toxic chemicals,

HORITSU, H., TAKADO, M. L. DAKA, E., TOMOYEDA, AND M. AND OGAWA, T. 1977. Degradation of p-aminoazobenzee by *Bacillus subtilis*. European Journal of Applied Microbiology. 4: 217-224.

HORVATH, R.S. 1972. Microbial co-metabolism and the degradation of organic compounds in nature. Bacteriol. Rev. 36:146-155.

HUGHES, M.N. AND POOLE, R.K., 1991. Metal speciation and microbial growth—the hard (and soft) facts. J. Gen. Microbiol. 137:725-734.

HULBERT, M.H. AND KRAWIES, S. 1977. Co-metabolism, a critique. J. Theor. Biol. 69:287-291.

HWANG, H.M., HODSON, R.E. AND LEE, R.F. 1987. Degradation of aniline and chloroaniline by sunlight and microbes in estuarine water. Water Research. 21(3):309-316.

IARC (International Agency for Research in Cancer). 1972-1990. Monographs on the Evaluation of Carcinogenic Risks to Humans, Vol. 1-49, Lyons, France.

JANI, D.K. 1997. M.Sc. dissertation thesis, Saurashtra University, Rajkot, India.

JANSEN, R.A. 1976. Enzymatic recruitment in evolution of new function. Annu. Rev. Microbiol. 30:409-425.

JEFFRIES, T.W 1990. Biodegradation of lignin-carbohydrate complexes. Biodegradation. 1:163-176.

JIMENEZ, L., BREEN, A., THOMAS, N., FEDERLE, T.W. AND SAYLER, G.S. 1991. Mineralization of LAS by a four-member aerobic bacterial consortium. Applied and Environmental Microbiology. 57(5):1566-1569.

JJANDRA SETIADI AND MARK VAN LOOSDRECHI, 1997. Anaerobic decolorization of textile wastewater containing reactive azodyes Proc. The 8th international conference on anaerobic digestion, Sendai, Japan, May 25-29, 1997.

KECK, A., KLEIN, J., KUDICH, M., STOLZ, A., KNACKMUSS, H.J. AND MATTES, R. 1997. Reduction of azo dyes by redox mediators originating in the naphalenesulfonic acid degradation pathway of Sphingomonas sp. Strain BN6. Applied and Environmental Microbiology. 63(9):3684-3690.

KELLY, I. AND CERNIGLIA, C.E. 1991. The metabolism of fluoranthrene by a species of Mycobacterium. J. Ind. Microbiol. 7:19-26.

KELLY, I., FREEMAN, J.P., EVANS, F. E. AND CERNIGLIA, C.E. 1993. Identification of metabolites from the degradation of metabolites from the degradation of fluoranthrene by Mycobacterium sp. strain PYR-1. Appl. Environ. Microbial. 59:800-806.

KEUTH, S. AND REHM, H.J. 1991. Biodegradation by phenanthrene by *Arthrobacter polychromogens*. Isolated from a contaminated soil. Appl Microbiol I Biotechnol. 34:804-808.

KIRK, T.K. 1981a. Principle of lignin degradation by white rot fungi in "The Ekman Days 1981. Int. Symp. Wood and Pulp. Chemistry, Stockholm, Sweden, Vol III. pp. 66-70.

KIRK, T.K. 1981b. Toward elucidating the mechanism of action of the ligninolytic systems in basidiomycetes. In "Trends in the Biology of Fermentation for Fuels and Chemicals", Hollander. A., R. Rabson and P. Rogers, A., Vol. III. pp. 66-70.

KIRK, T.K., AND CHANG, H.M. 1981. Potential applications of bio-ligninolytic systems. Enzyme Microb. Technol. 3:189-196.

KIRK, T.K., AND FARRELL, R.L. 1987. Enzymatic "combustion": The microbial degradation of lignin. Annu. Rev. Microbiol. 41:465-505.

KIRK, T.K., S. CROAN, M. TIEN, K.E. MURTAGH AND FARRELL, R.L. 1986. Production of multiple ligninases by *Phanerochaete chrysosporium* effect of selected growth conditions and use of mutant strain. Enzyme Microb Technol. 8:27-32.

KONOPKA, A. 1993. Isolation and characterization of a subsurface bacterium that degrades aniline and methylanilines. FEMS Microbiology Letters. 111(1):93-99.

KUDLICH, M., KECK, A., KLEIN, J. AND STOLZ, A. 1997. Localization of the enzyme system involved in anaerobic reduction of azo dyes by *Sphingomonas sp.* Strain BN6 and effect of artificial redox mediators on the rate of azo dye reduction. Applied and Environmental Microbiology. 63(9):3691-3694.

KULLA, H.G. 1981. Aerobic bacterial degradation of azo dyes. In: Microbial degradation of xenobiotics and recalcitrant compounds, Leisinger T., Cook, A.M., Hutter, R. and Nuesch, J., Academic Press, London. pp. 387-399.

KULLA, H.G., KLAUSENER, F.K., MEYER, U., LUDEKE, B. AND LEISINGER, T. 1983. Interference of Orange II. Archives of Microbiology. 135:1-7.

KUREK, B. AND MONTIES, B. 1994. Oxidation of spruce lignin by fungal lignin peroxidase and horseradish peroxidase: Comparison of their actions on molecular structure of the polymer in colloidal solutions. Enzyme Microb. Technol. 16:125-130.

LANDRUM, P.F., W.S. DUPUIS AND KUKKONEN, J. 1994. Toxicokinetics and toxicity of sediment-associated pyrene and phenanthrene in Diporeia spp.: examination of equilibrium partitioningteoru and residue based effects for assessing hazard. Environ. Toxicol. Chem. 13:1769-1780.

LEONOWICZ, A., MATUSZEWSKA, A., LUTEREK, J., ZIEGENHAGEN, D., WOJTAS-WASILEWSKA, M., CHO N.S., HOFRICHTER, M. AND ROGALSKI, J., 1999. Biodegradation of lignin by white-rot fungi. Fungal Genet Biol. 27:175-185.

LEVINE, W.G. 1991. Metabolism of azo dyes: implications for detoxification and activation. Drug Metabolism Reviews. 23 (3-4):253-309.

LIE T.J., LEADBETTER, J.R. AND LEADBETTER, E.R. 1998. Metabolism of sulfonic acids and other organosulfur compounds by sulfate-reducing bacteria. Geomicrobiology Journal. 15(2):135-149.

LIN, Z. AND YANG, H. 1989. The decolorization and biodegradation of azo dyes by Pseudomonas S-42. Acta Microbiology Sinica. 29:418-426.

LIN, Z. AND YANG, H. 1991. Decolorization and biodegradation metabolism of azo dyes by Pseudomonas S-42. Journal of Environmental Science. 3:89-102.

LOCHER, H.H., THUMHEER, T., LESINGER, T. AND COOK, A.M. 1989. 3-Nitrobenzesulfonate, 3-aminobenzenesulfonate, and 4-aminobenzenesulfonate as sole carbon sources for bacteria. Applied and Environmental Microbiology. 55(2):492-494.

LOIDI, M., HINTEREGGER, C., DITZELMULLER, G., FERSCHL, A. AND STREICHSBIER, F. 1990. Degradation of aniline and monochlorinated anilines by soil-born Pseudomonas acidovorans strains. Archieves of Microbiology. 155(1):56-61.

LYONS, C.D., KATZ, S. AND BARTHA, R. 1984. Mechanisms and pathways of aniline elimination from aquatic environments. Applied and Environmental Microbiology. 48(3):491-496.

MAHAFFEY, W.R., GIBSON, D.T. AND CERNIGLIA, C.E. 1988. Bacterial oxidation of chemical carcinogens: Formation of polycyclic aromatic acids from Benz(a)anthracene. Appl. Environ. Microbial. 54:2415-2423.

MAHNOT, S.M. AND BHANDARI, M.M. 1987. Environmental Degradation in western Rajasthan. Jodhpur University Press, Jodhpur p. 136

MAMPEL, J., HITZLER, T., RITTER, A. AND COOK, A.M. 1998. Desulfonation of biotransformation products from commercial linear alkylbenzenesulfonates. Environmental Toxicology and Chemistry. 17(10):1960-1963.

MANNING, B.W. CERNIGLIA, C.F. AND FEDERLE, T.W. 1985. Metabolism of the benzidine - based azo dye Direct blue Black-38 by human intestinal microbiota. Applied and Environmental Microbiology. 50: 10-15.

MAYER, U. 1981. Biodegradation of synthetic organic colorants. In: Microbial degradation of xenobiotics and recalcitrant compounds. T. Leisinger, R. Hutter, A.M. Cook and J. Nuesch. Eds. pp. 387-399. London: Academic Press.

MCKENNA, E. 1979. Biodegradation of Polynuclear Aromatic Hydrocarbon Pollutants by Soil and Water Microorganisms, Final Report, Projects No. A-073-ILL, University of Illinois, Water Resources Centre, Urbana, IL,

MECHSNER, K. AND WUHRMANN, K. 1982. Cell permeability as a rate-limiting factor in the microbial reduction of sulfonated azo dyes. European Journal of Applied Microbiology and Biotechnology. 15:123-126.

MICHAELS, G.B. AND LEWIS, D.L. 1985. Sorption and toxicity of azo ad triphenylmethane dyes to aquatic microbial populations. Environ. Toxicol. Chem. 4:45-50.

MONNA, L., OMORI, T. AND KODAMA, T. 1991. Microbial degradation of dibenzofuran, Fluorene, and dibenzo-p-dioxin by *Staphylococcus auriculans* DBF63. Appl. Environ. Microbial. 57:1441-1447.

MULLER, J.G., CHAMPMAN, P.J., BLATTMAN, B.O. AND PRITCHARD, P.H. 1990. Isolation and characterization of a fluoranthrene–utilizing strain of *Pseudomonas paucimobilis*. Appl. Environ. Microbial. 56:1079-1086.

NIGAM, P., BANAT, I.M., SINGH, D. AND MARCHANT, R. 1996a. Microbial process for the decolorization of textile effluent containing azo, diazo and reactive dyes. Process Biochemistry. 31(5):435-442.

NIGAM, P., MCMULLAN, G., BANAT, I.M. AND MARCHANT, R. 1996b. Decolourization of effluent from the textile industry by a microbial consortium. Biotechnology Letters. 18(1):117-120.

OLLIKKA, P. ALHONMAKI, K., LEPPANEN, V.M., GLUMOFF, T., RAIJOLA, T. AND SUOMINEN, I., 1993. Decolorization of azo, triphenylmethane, heterocyclic, and polymeric dyes by lignin peroxidase isoenzymes from *Phanerochaete chrysosporium*. Appl. Environ. Microbiol. 59:4010-4016.

ORGANIZATION FOR ECONOMIC CO-OPERATION AND DEVELOPMENT, (OECD) 1994. Biotechnology for a clean Environment: Prevention, detection, and remediation, OECD. Paris.

ORTH, A.B. AND TIEN, M. 1995. Biotechnology of lignin degradation. In: Esser K., Lemke, P.A. (eds.) The Mycota II. Genetics and biotechnology. Springer, Berlin Heidelberg New York. pp. 287-302.

PAGGA, U. AND BROWN, D., 1986. The degradation of dyestuffs: Part II Behavior of dyestuffs in aerobic biodegradation tests. Chemosphere. 15(4):479-491.

PALMER, J.M., HARVEY, P.J. AND SCHOEMAKER, H.E. 1986. In: Utilization of lingo cellulosic wastes, (ed.) Hartley, B.S., Broda, P.M.A. and Senior. P.J. London: The Royal Society.

PASTI, M.B. AND D.L. CRAWFORD, 1991. Relationship between the abilities of streptomycetes to decolorize three anthrone type dyes and to degrade lignocellulose. Can. J. Microbiol. 37:902-907.

Pasti-Grigsby, M.B., Paszcynski, A., Gosczynski, S., Crawford, D.L. and Crawford, R. L. 1992. Influence of aromatic substitution patterns on azo dye degradability by *Streptomyces sp.* and *Phanerochaete chrysosporium*. Appl. Environ. Microbiol. 58:3605-3613.

Paszczynski, A., M.B. Pasti, S. Goszczynski, D.L. Crawford, and Crawford, R. L. 1991. New approach to improve degradation of recalcitrant azo dyes by *Streptomyces spp* and *Phanerochaete chrysosporium*. Enzyme Microb. Technol. 13:378-384.

Paszczynski, A., M.B. Pasti-Grisby, S. Goszczynski, R.L. Crawford and Crawford, D.L. 1992. Mineralization of sulfonayted azo dyes and sulfanilic acid by *Phanerochaete chrysosporium* and *Streptomyces chromofuscus*. Appl. Environ. Microbiol. 58:3598-3604.

Patil, S.S. and Shinde, V.M. 1988. Biodegradation studies of aniline and nitrobenzene in aniline plant wastewater by gas chromatography. Environmental Science and Technology. 22(10):1160-1165.

Pointing, S.B. 2001. Feasibility of bioremediation by white rot fungi. Mini review. Applied microbial biotechnology. 57:20-33.

Pointing, S.B. Bucher, V.V.C. and Vrijmoed, L.L.P. 2000b. Dye decolorization by sub-tropical basidiomycetous fungi and the effect of metals on decolorizing ability. World J. Microbiol. Biotechnol. 16:199-205.

Pointing, S.B. and Vrijmoed, L.L.P. 2000. Decolorization of azo and triphenylmethane dyes by *Pychoporus sanguineus. Sanguineus* producing laccase as the sole phenoloxidase. World J. Microbiol. Biotechnol. 16:317-318.

Rafii, F., Franklin, W. and Cerniglia, C.E. 1990. Azo reductase activity of anaerobic bacteria isolated from human intestinal microflora. Applied and Environmental Microbiology. 56(7):2146-2151.

RAINA, M., MAIER, LAN. L. PEPPER AND CHARLES P. GERBA. 2000. Environmental Microbiology, Academic Press. pp. 363-364.

RAJANNAN, G. AND OBLISAMI, G. 1979. Effect of paper factory effluents on soil and crop plants. Indian J Environ. Health. 21:120-130.

RAZO-FLORES, E., LUITEN, M., DONLON, B. A., LETTINGA, G. AND FIELD J. A. 1997. Complete biodegradation of the azo dye azodisalicylate under anaerobic conditions. Environmental Science and Technology. 31(7):2089-2103.

REBER, H., HELM, V. AND KARANTH, N. G. K. 1979. Comparative studies on the metabolism of aniline and chloroaniline by *Psuedomonas multivorans* strain an European Journal of Applied Microbiology and Biotechnology. 7:181-189.

REDDY, C. A. AND D'SOUZA, T.M., 1994. Physiology and molecular biology of the lignin peroxidases of Phanerochaete chrysosporium. FEMS Microbiol. Rev. 13:137-152.

REDDY, C.A. 1995. The potential for white-rot fungi in the treatment of pollutants. Curr. Opin. Biotechnol. 6:320-328.

REID, T.M., MORTON, K.C., WANG, C.Y. AND KING C.M. 1984. Mutagenicity of azo dyes following metabolism by different reductive/oxidative systems. Environmental Mutagenicity, 6:705-717.

RIPIN, M.J., NOON, K.F. AND COOK, T.M. 1971. Bacterial Metabolism of arylsulfonates. Applied Microbiology. 21(3):495-499.

RODIRIGUEZ, E., PICKARD, M.A. AND VASQUEZ-DUHALT, R. 1999. Industrial dye decolorization by laccases from ligninolytic fungi. Curr. Microbiol. 38:27-32.

ROSENKRANZ, H.S. AND KLOPMAN, G. 1990. Structural basis of the mutagenicity of 1-amino-2-naphthol-based azo dyes. Mutagenesis, 5(2):137-146.

ROSENKRANZ, H.S. AND KLOPMAN, G. 1989. Structural basis of the mutagenicity of phenylazoaniline dyes. Mutation Research. 221(3):217-234.

RUSS, R., MUELLER, C., KNACKMUSS, H.J. AND STOLZ, A. 1994. Aerobic biodegradation of 3-aminobenzoate by Gram-negative bacteria involves intermediate formation of 5-aminosalicylate as ring-cleavage substrate. FEMS Microbiology Letters. 122 (1-2):137-143.

SAKAKIBARA, A. 1983. Chemical structure of lignin related mainly to degradation products. In: Recent advances in lignin biodegradation research. T. Higuchi, H.-M.Chang and T.K.Kirk (eds.) UNI, Tokyo, pp. 12-33.

SASEK, V., O. *Volfova*, P. *Erbanova*, B.R.M. *Vyas* & *Matucha*, M. 1993. Biotechnol. Lett. 15:512-526.

SCHLEGEL, H.G. AND H.W. JANNASCH. 1992. Prokaryotes and their habitats, pp. 43-82. In: A. Balows, H.G. Truper, M. Dworkin, W. Harder, and K.H. Schleifer (ed), The prokaryotes, 2nd ed. Springer Verlag, New York.

SCHOEMAKER, H.E. HARVEY, P. J., BOWEN, R.M. AND PALMER, J. M. 1985. FEBS Letters. 83:7-12.

SHAUL, G.M., HOLDSWORTH, T. J., DEMMPSEY, C.R. AND DOSTAL, K. A. 1991. Fate of water-soluble azo dyes in the activated sludge process. Chemosphere. 22:107-119.

SHELTON, D.R. AND J. M. TIEDJE 1994. Isolation and partial characterization of bacteria in an anaerobic consortium that mineralize 3- chlorobenzoic acid. Appl. Environ. Microbiol. 48:840-848

SHIARIS, M.P. 1989. Seasonal Biotransformation of Naphthalene, Phenanthrene, and Benzo[a]Pyrene in Superficial Estuarine Sediments. Appl. Environ. Microbiol. 55:1391-1399.

SLOBE, J.F. AND DE L.G.E. 1986. Effects of land use on fresh waters agriculture, forestry, mineral exploitation, and sanitation. Ellis Horwood Limited, Chishester, U.K. p. 568.

SPADARO, J.T., GOLD AND M.H, RENGANATHAN, V. 1992. Degradation of azo dyes by the lignin-degrading fungus Phanerochaete chrysosporium. Appl. Environ. Microbiol. 58:2397-2401.

STALEY, T. JAMES, MARVIN P. BRYANT, NORBERT PFENNING AND JOHN G. HOLT. 1984. Bergey,e Mannual of Systematic Bacteriology.

STANIER. R.Y. AND N.J. PALLERONI, AND M. DOUDOROFF, 1966. The aerobic Pseudomonas: A taxonomic study. J. Gen. Microbiol. 43:159-271.

STOLZ, A., NORTEMANN, B. AND KNACKMUSS, H.J. 1992. Bacterial metabolism of 5-aminosalicylic acid: Initial ring cleavage. Biochemical Journal. 282(3):675-680.

STONER, D.L. 1994. Biotechnology for the treatment of hazardous waste. Baca Raton, FL: CRC Press.

STRINGFELLOW, W.T. AND M.D. AITKEN 1994. Comparative physiology of phenanthrene degradation by two dissimilar Pseudomonas isolated from a creosote-contaminated soil. Can. J. Microbiol. 40:432-438.

SUTHERLAND, B., F. RAFIC, A.A. KHAN, AND C.E. CERNIGLIA 1995. Mechanism of polycyclic aromatic hydrocarbon degradation, pp. 269-300. In: L.Y. Young and C.E. Cerniglia (ed.), Microbial Transformation and Degradation of toxic Organic Chemicals. Wiley-Liss, New York.

SUTHERSAN, S. S. 1999. "In Situ Bioremediation" Remediation engineering: Design Concept, Boca Raton: CRC Press LLC.

TAN, N.C.G. AND FIELD, J.A. 2000. Biodegradation of sulphonated aromatic compounds. In: Environmental technologies to treat sulfur pollution. Principles and engineering, Lens P. and Hulshoff pol. L. IWA Publishing, London. pp. 377-392.

THOUMELIN, G. 1991. Presence and behavior of linear alkylbenzenesulfonates (LAS) and alkyl phenol ethoxylates (APE) in the aquatic environment: Review. Environmental Technology. 12(11):1037-1046.

THURSTON, C.F. 1994. The structure and function of fungal laccases. Microbiology.140:19-26.

TIEN, M. AND KIRK, T. K. 1983. Science. 221:661-663.

TIEN, M. AND KIRK, T. K. 1984. Proceedings of the Natural Academy of Science of the U.S.A. 81:2280-2284.

TIEN, M. AND T. K. KIRK 1988. Lignin peroxidase of Phanerochaete chrysosporium Methods Enzynol. 161:238-249.

TJANDRA SETIAD AND MARK VAN LOODRECHI, Anaerobic decolorization of textile wastewater containing Reactive azo dyes. 1997. Proc. The 8 th International Conference on Anaerobic Digestion, Sendai, Japan, May 25-29.

VAIDYA, A.A. AND DATYE, K.V. 1982. Environmental pollution during chemical processing of synthetic fibres. Colourage. 14:3-10.

VAN DER OOST, R., F.J. VAN SCHOOTEN, F. ARIESE AND H. HEIDA 1994. Bioaccumalation, biotransformation and DNA binding of PAHs in feral eel (Anguilla anguilla) exposed to polluted sediments: a field survey. Environ. Toxicol. Chem. 13:859-870.

VAN DER ZEE F.P., LETTINGA, G. AND FIELD, J.A. 2000b. Azo dye decolorization by anaerobic granular sludge. Chemosphere (accepted).

VAN DER ZEE F.P., BOUWMAN, R.H.M., STRIK D.P.B.T.B., LETTINGA, G. AND FIELD, J. A. 2002a. Application of redox mediators to accelerate the transformation of reactive azo dyes in anaerobic bioreactors. Biotechnology and Bioengineering.

VAN DER ZEE, F. P., LETTINGA, G. AND FIELD, J. A. 2000C. The role of (auto) catalysis in the mechanism of anaerobic azo reduction. Water Science and Technology. 42(5-6):301-308.

VYAS, B. R. M. AND H. P. MOLITORIS. 1995. Involvement of an extracellular H_2O_2-dependent ligninolytic activity of the white rot fungus Pleurotus ostreatus in the decolorization of Remazol Brilliant Blue R. Appl. Environ. Microbiol. 61:3919-3927.

VYAS, B.R., M.J. VOLC, AND V. SASEK. 1994a. Ligninolytic enzymes of selected white rot fungi cultivated on wheat straw. Folia Microbiol. 39:235-240.

WALKER, R. 1970.The metabolism of azo compounds: a review of the literature. Food and Cosmetic Toxicology. 8:659-676.

WALKER, R. AND RYAN, A.J. 1971. Some molecular parameters influencing rate of reductions of azo compounds by intestinal microflora. Xenobiotica. 1(14/5):483-486.

WALTER, U., BEYER, M., KLEIN J. AND REHM, H.J. 1991. Degradation of pyrene by Rhodococcus sp. UW1. Appl. Microbiol. Biotechnol. 34:671-676.

WARIISHI, H., VALLI, K. AND GOLD, M.H. 1992. Manganese (II) oxidation by manganese peroxidase from the basidiomycete Phanerochaete chrysosporium. J. Biol. Chem. 267:23688-23695.

WEBER, E.J. AND WOLF, L.N. 1987. Kinetic studies of the reduction of aromatic azo compound in anaerobic sediment/ water systems. Environmental Toxicology and Chemistry. 6:911-919.

WEISSENFELS, W.D., BEYER, M. AND KLEIN, J. 1990. Degradation of phenanthrene, fluorene and fluoranthrene by pure bacteria cultures. Appl. Microbiol. Bitechnol. 32:479-484.